THE EARLY DAYS
OF THE
POWER STATION INDUSTRY

THE EARLY DAYS
OF THE
POWER STATION INDUSTRY

by

R. H. PARSONS, M.I.Mech.E., M.E.I.C.

Author of
"The Development of the Parsons Steam Turbine"

CAMBRIDGE
AT THE UNIVERSITY PRESS
1940

CAMBRIDGE
UNIVERSITY PRESS

University Printing House, Cambridge CB2 8BS, United Kingdom

Cambridge University Press is part of the University of Cambridge.

It furthers the University's mission by disseminating knowledge in the pursuit of education, learning and research at the highest international levels of excellence.

www.cambridge.org
Information on this title: www.cambridge.org/9781107475045

© Cambridge University Press 1940

First published 1940
First paperback edition 2014

A catalogue record for this publication is available from the British Library

ISBN 978-1-107-47504-5 Paperback

CONTENTS

LIST OF PLATES

DIAGRAMS IN TEXT

PREFACE

THE supply of electricity as a Public Service is a comparatively new industry, as it has only been in existence for less than sixty years, yet so rapid has been its growth and so profound the changes it has undergone in that time, that hardly any of its original features are now recognizable. It therefore seemed worth while, as a matter of historical interest, to give some account of electricity supply as it was in the days when every enterprise was largely of a pioneering nature, and the future of the industry was still unpredictable. This period may be said to have closed at the end of the last century, for by that time the lines along which progress would take place were becoming evident. The steam turbine had already foreshadowed the doom of reciprocating machinery in Power Stations, and the three-phase turbo-alternator producing high-voltage current for conversion or transformation in substations was clearly indicated as the generating unit on which Central Station practice would be based.

Developments subsequent to this period, although of the highest technical interest and importance, fall outside the scope of the present book. The aim of the author has been to show how the industry originated, to describe various systems of generation and distribution adopted by particular undertakings, and to give some idea of the kind of machinery with which the early Power Stations were equipped. Accuracy in detail has been ensured as far as possible by systematic reference to contemporary records, and in this connection the information gathered from the old volumes of the *Electrical Review, Electrician, Engineer, Garcke's Manual of Electrical Undertakings* and other publications has been made the fullest use of. Acknowledgment must also be made to a series of interesting articles on the early history of certain London undertakings by Mr C. Kibblewhite which appeared last year in *Contact*, the staff journal of Central London Electricity, Ltd.

The author desires to express his particular thanks for assistance and information so readily given by Mr J. G. Freeman, late

Chief Engineer of the London Electric Supply Corporation, who could verify from personal knowledge the facts concerning the history of the famous Grosvenor Gallery and Deptford Stations; by Mr H. P. Gaze, Chief Generating Engineer of the London Power Company, whose experience dates from the Mason's Yard Station of the St James' and Pall Mall Electric Lighting Company; by Mr E. Harlow, the Chief Engineer of the City of London Electric Lighting Company, who permitted access to records going back to the earliest days of the Company; by Mr C. A. Holbrow of Messrs C. A. Parsons and Co., Ltd., who was one of the erectors and subsequently one of the engineers of the Gordon Station at Paddington in 1886; by Mr F. D. Napier of Messrs Babcock and Wilcox, whose Central Station experience covers the operation of the Whitehall Court and Amberley Road Stations; by Mr Roger Smith, of Messrs Highfield and Roger Smith, who was formerly Chief Electrical Engineer to the Great Western Railway Company; and by Mr Percy Still, the Chief Engineer of the Chelsea Electricity Supply Company from 1892 until his retirement in 1937. Thanks are also due to Messrs Pritchett and Gold and E.P.S. Company, Ltd., who, as the successors of the Electrical Power Storage Company, Ltd., were most helpful in respect of the work carried out by the latter Company for the Chelsea and other undertakings about fifty years ago.

"And here will I make an end.

If I have done well, and as is fitting the story, it is that which I desired: but if slenderly and meanly, it is that which I could attain unto."

R. H. P.

1939

CHAPTER I

THE BEGINNINGS OF THE POWER STATION INDUSTRY

> They had no vision amazing
> Of the goodly house they are raising,
> They had no divine foreshowing
> Of the land to which they are going;
> But on one man's soul it hath broken,
> A light that doth not depart;
> And his look, or a word he hath spoken,
> Wrought flame in another man's heart.
>
> A. W. E. O'SHAUGNESSY

THE birth of the Power Station Industry was foreshadowed by Faraday's discovery, in 1831, that electricity could be generated by mechanical means, though many years had to elapse before machinery was available for its production on a commercial scale, or apparatus for its utilization had been devised. The development of the dynamo in its present form may be said to have started in 1845 when Wheatstone and Cooke patented the use of electro-magnets in place of permanent magnets for the field. Machines were not, however, made completely self-exciting before 1866 when the brothers C. and S. A. Varley, Dr Werner Siemens and Sir Charles Wheatstone, independently and practically simultaneously, discovered the principle of self-excitation. The production by Gramme, in 1870, of a dynamo with a ring-wound armature then brought matters to a point when the industrial generation of electricity really became a practical question. The following years were prolific in inventions concerning dynamo-electric machinery, notable amongst these being the open-coil dynamo of Brush in 1878, which played a prominent part in the early history of electric lighting. Long before this time, however, serious attempts had been made to obtain electricity from mechanical power by the use of machines of the magneto type.

The first Company to be formed for the exploitation of electric machinery was probably La Société Générale d'Electricité, of

Paris, which was founded in 1853 to develop a machine designed by Professor Nollet for the purpose of generating current for the electrolysis of water. The object was to produce hydrogen and oxygen for making limelight. The project was not successful, but the machine was modified by Professor F. H. Holmes and used for experiments in connection with electric light. Holmes's first machine was tried in the lighthouse at Blackwall in 1857, and on December 8 of the next year, light produced by his second machine was thrown over the sea from the South Foreland Lighthouse. These experiments were so promising that about 1859 another Company, called the Compagnie de l'Alliance, was formed for the manufacture of electric generators of the type in question. The "Alliance" machines, as they were called, produced alternating current, and were used to a considerable extent for lighthouse work by both the French and British authorities. Their characteristic feature was the production of their magnetic fields by means of a large number of permanent magnets of horseshoe shape.

On 1 February 1862, electric light was installed permanently in the Dungeness Lighthouse, the machine and lamp being of Holmes's design. A Holmes machine, dating from 1867, which was installed in the Souter Point Lighthouse in 1871, is preserved by the institution of Electrical Engineers, to whom it was presented by the Corporation of Trinity House on its removal from the lighthouse in 1915. The machine, which weighs 3 tons, was driven at 400 R.P.M. by an Allen engine. It absorbed 32 H.P., from which a light of 1,520 c.p. was obtained. Alternators of the same type were employed for operating the two lighthouses at Cap la Hêve, near Havre, in 1863 and 1865 respectively.

The fact that a brilliant light could be produced by allowing an electric current to form an arc between a pair of carbon points had been demonstrated by Sir Humphry Davy before the Royal Institution in 1808, the current being obtained from a battery of 2,000 zinc-copper cells. Arc-lighting was for many years the only way of utilizing electricity for illumination. A mechanical arc lamp was produced in 1847 by Staite, who was followed by numerous other inventors, but the first type of arc lamp to achieve success on a large scale was the famous "Jablochkoff Candle" invented in 1876. This consisted essentially of two carbon rods placed parallel to each other and separated by a plate of kaolin. A bridge of carbon paste connecting the tips of the rods was burned away when the current was switched on, and the arc thus formed maintained itself between the carbon rods, volatilizing the intervening partition as the rods were consumed. Self-regulating arc lamps did not appear until 1878, when von Hefner-Alteneck introduced the differential solenoid type, and Brush devised the clutch mechanism to effect the same purpose.

With dynamos and arc lamps at their disposal, the pioneers of the electric supply industry were able to proceed. They started, as we have seen, with lighthouse installations, as the intensity of the light rendered the arc lamp specially suitable for such work. On land the only field for the new light was in the illumination of large spaces, and one of the first applications for such a purpose was the lighting of the Gare du Nord in Paris in 1875. In May 1877 the Grands Magasins du Louvre, in the same city, put down an installation of 80 Jablochkoff candles supplied with current from Gramme machines driven by a steam engine in the basement. Within the next 18 months, this system of lighting had greatly extended. In 1878 there were several hundred Gramme machines in service, the latest type supplying alternating current, as this was found more suitable for Jablochkoff candles on account of the equal consumption of the two carbons. These machines had eight salient rotating field poles and a fixed armature. The largest absorbed about 16 H.P. at 600 R.P.M. and would serve 16 Jablochkoff candles. The Avenue de l'Opéra was lit by 46 lamps supplied from three 20 H.P. engines in different places; the Place de l'Opéra had 22 lamps, and there were altogether over 300 Jablochkoff candles in Paris.

Progress in England was slower. In its issue of 20 July 1878, *The Electrician* bewailed the fact that although the use of the electric light was daily extending in Paris, "yet in London there is not one such light to be seen". Londoners, however, had not very much longer to wait, for in the following month six Lontin lamps, installed by French engineers, were employed to illuminate the Gaiety Theatre, this being the first public building in

London to be electrically lighted. The effect was described at the time as that of "half a dozen harvest moons shining at once in the Strand". The Jablochkoff system, so successful in Paris, was introduced in this country on October 15 of the same year by Messrs Wells and Co., of the Commercial Iron Works, Shoreditch, who put up four lamps inside their showroom and two more at the entrance to their works, current being furnished by a Gramme machine.

The first move of importance by a Public Authority with regard to electric light was made on 18 October 1878, when, on the recommendation of their Works and General Purposes Committee, the old Metropolitan Board of Works accepted an offer of the Société Générale d'Electricité of Paris to instal an experimental system of lighting along the Thames Embankment. The Company were to supply 20 Jablochkoff lamps and the necessary electrical machinery, while the Board would bear the cost of providing the motive power, cables, standards, labour, etc. Within a week of this decision the City of London Authorities came to an agreement with the same Company to try the electric light along the Holborn Viaduct and in front of the Mansion House. Before, however, either of these schemes could be realized, the Billingsgate Fish Market, controlled by the Markets Committee of the City Corporation, was lit both inside and outside by electricity. The inauguration of the new system took place on 29 November 1878, thus giving Billingsgate the credit of affording the first demonstration of electric lighting by any Public Authority in London. The installation, which comprised 16 Jablochkoff candles in opaline globes, was also carried out by the Société Générale d'Electricité.

The Victoria Embankment was illuminated by electricity for the first time on 13 December 1878. The lights, which were along the river wall of the Embankment between the Waterloo and Westminster bridges, consisted of 20 Jablochkoff lamps spaced about 45 yards apart. On the other side of the Embankment, just west of Charing Cross Bridge, was a wooden shed containing a semi-portable steam engine constructed by Messrs Ransome, Sims and Head, having two cylinders each 10 in. diameter by

13 in. stroke. The engine, which worked with steam at 60 lb. pressure, was capable of developing about 60 I.H.P. at 160 R.P.M. It was belted to a countershaft from which were driven a direct current Gramme dynamo at 650 R.P.M. and its separate exciting dynamo at 700 R.P.M. The lamps were arranged five in series on four circuits, the eight conductors of bare wire being led through a 4 in. drain-pipe to the subway and thence to the lamp standards. The farthest lamp in the direction of Waterloo Bridge was 470 yards from the engine-house, and the farthest in the other direction was at a distance of 700 yards from the house. The original engine, after running nearly 5 years, was replaced by a Davey Paxman engine in August 1883. The system was extended on 16 March 1879, to a total of 40 lamps, and on October 10 following, a further extension brought the number of lamps to 55, the mains then extending from a point 6,092 ft. below Waterloo Bridge to a point 6,007 ft. above it. In June 1881 an agreement was entered into between the Metropolitan Board of Works and the Jablochkoff Electric Light and Power Co., by which the Company should maintain 40 lights on the Embankment and 10 on Waterloo Bridge for the price of $1\frac{1}{2}d.$ per lamp-hour. After the termination of this arrangement in 1884, the Company went into liquidation, as the price had been an unremunerative one, and as no other Company could be found willing to undertake the work on terms satisfactory to the Board, gas lighting was then reverted to.

The Holborn Viaduct installation was put into commission before the end of 1878, but the lighting of the Viaduct by electricity was discontinued on 9 May 1879, as being too costly, the expense being stated to be about four times that of adequate lighting by gas. It comprised 16 Jablochkoff lamps distributed over a distance of 473 yards, and supplied with current from a Gramme dynamo driven by a Robey undertype engine, the details of the work being generally similar to those of the undertaking on the Embankment.

In addition to the installations mentioned, the year 1878 also witnessed much enterprise by industrial undertakings and private individuals with regard to electric lighting, and various municipalities also commenced to take an interest in the subject. Before the close of the year *The Times* was using the new light in its printing office, and it was also being employed in such establishments as Messrs Pullar's Dye Works at Perth, the Steel Works of Messrs Cammell and Wilson at Dronfield, and Messrs Shoolbred in London. In Woolwich Arsenal it was installed in various departments, the Trafalgar Colliery in the Forest of Dean was using it for pit-head lighting, the London, Brighton and South Coast Railway had electric light in their London Bridge terminus, and St Enoch's Station in Glasgow was lit by Crompton arc lamps. Electricity had been tried for street lighting at Westgate-on-Sea, a game of football had been played by electric light before 30,000 spectators at Sheffield, and Sir William Armstrong had put down a small hydro-electric plant to generate current for the lighting of his picture gallery at Craigside, a mile and a quarter away.

An even greater activity prevailed during the year 1879. Messrs W. D. and H. O. Wills put down an electric light plant for their tobacco factory at Bristol under the advice of Professor S. P. Thompson who was then at the Bristol University. The Reform Club, the Langham Hotel, St George's Pier Head at Liverpool, the Avonmouth Docks, the sea front at Blackpool, etc., were all provided with the new form of illumination, which was also installed in the Reading Room of the British Museum, while the constructional work on the great railway bridge across the Severn at Lydney was also much facilitated by the electric light.

The generating units most commonly employed in those days were the Gramme and Siemens machines belt-driven from some simple kind of steam engine which was almost invariably combined with its own boiler. The makers of agricultural machinery seem first to have appreciated the field for electric light engines. The power for the Embankment lighting was furnished, as already mentioned, by a semi-portable engine constructed by Messrs Ransome, Sims and Head; that for the lighting of the Holborn Viaduct and the Billingsgate Market by Robey engines; Messrs Wallis and Steevens of Basingstoke supplied the semi-portable engine for the British Museum lighting; while Messrs

Clayton and Shuttleworth's portable engines were employed for the lighting of the Thorncliffe Works of Messrs Newton Chambers and Co., and a Garrett portable engine for the lighting of the front at Westgate-on-Sea. Messrs Marshall, Sons and Co. of Gainsborough exhibited a portable engine provided with a wheeled forecarriage on which was a dynamo, at the Kilburn Show of the R.A.S.E. in 1879, and indeed hardly a maker of agricultural engines seems to have neglected the opportunity for business afforded by the coming of the electric light. The earliest attempt to produce a high-speed direct-coupled unit was made by Mr Peter Brotherhood who arranged his recently invented three-cylinder engine to drive a dynamo directly from either end of the crank shaft. A machine of this kind was used for the experimental lighting of the terminus of the P.L.M. Railway in Paris by means of Lontin arc lamps on 7 September 1877, and according to an advertisement that appeared in 1879, dynamos by Wilde, Siemens and Gramme had all been driven directly by Brotherhood engines.

When no convenient cellar or other place existed for the accommodation of generating machinery, the latter was often housed in a wooden building which was the prototype of the power station of to-day. The nature of one of these early installations is well described in a paper entitled "Three Months' Experience of Electric Lighting at Blackpool", read in 1880 by Mr W. Chew before the Manchester Institution of Gas Engineers. The plant in question was put down for the illumination of the sea front by means of six arc lamps, and went into commission on 18 September 1879. The station consisted of a substantial timber building 60 ft. long by 25 ft. wide. On the ground inside was a wooden framing which formed the foundation for the machinery, the interspaces being filled with concrete and then boarded over to prevent dust rising. Upon this were fixed two Robey portable engines, each of 16 H.P., and at the opposite end were seven Siemens dynamos all driven by belts from a countershaft overhead between them and the engines. To quote Mr Chew's own words:

From each of the six machines a distinct wire is led to its own lamp, and one common return wire so to speak, answers for the whole. You will no doubt wonder what the additional machine was for. Well, at first the whole of the six machines had to perform the duty of producing their own electricity, but that duty is now taken from them by specially devoting a dynamo electric machine to the purpose solely, and this was found to be better for the lights.

The "wires", it may be mentioned, were first laid underground in 2 in. and 3 in. cast-iron pipes, but sea water got in the trenches while the work was proceeding and destroyed the insulation. The whole length of 1500 ft. had therefore to be withdrawn and carried on poles overhead.

As regards the operation of the plant, Mr Chew says:

One engine man and his assistant look after the engines and boilers which require very great attention in their stoking, for a variation of 5 lb. pressure of steam will throw all the lights out of order. Another man is kept to oil the machines and keep all going right in that department as well as to look out for any light going out, which in some cases instantly shows itself at the machines, but always on a dial-board on which all the connections are made and on which we have suitable tell-tale apparatus.

The "dial-board" is clearly the original form of the modern switchboard, but it would be interesting to know the nature of the "suitable tell-tale apparatus". In a priced schedule of the equipment of the station, we find a "Strophometer" costing £10, a "Velocimeter" costing £2, and a "Galvanometer battery and Leclanché" priced at £5, these items apparently covering the whole of the instruments employed.

In October 1878 it was announced by Edison that he had "solved the problem" of "subdividing the electric light", the solution being, of course, the employment of incandescent lamps in parallel. To show how firmly the arc lamp was then established, in the minds, even of the best informed authorities, as the only practical source of electric light, it is interesting to recall the words of so famous an engineer as Professor Sylvanus P. Thompson, uttered in the course of a lecture on "The Electric Light" which he delivered on November 8 in the Colston Hall at Bristol. Referring to the recently reported announcement of Edison, Professor Thompson said:

I cannot tell you what Mr Edison's particular method of distributing the current to the spirals may be, but this I can tell you as the result of all experience, that any system of lighting by incandescence will utterly fail from an economic point of view, and will be the more uneconomical the more the light is subdivided.

The general disbelief that incandescent lamps would ever be able to displace arc lamps, or even to come into use on a practical scale for any kind of lighting, was also illustrated by the absence of any direct reference to incandescent lighting in the evidence given by such men as Sir William Thomson, Dr Hopkinson, Siemens and others before the Select Committee of the House of Commons in 1879, when an enquiry was being held in view of legislation concerning the new form of illumination.

Edison's first experiments with incandescent lamps were made with filaments of the rare metals, and it was not till later that he resorted to the use of filaments of carbon. Meanwhile, Mr Joseph Swan, who had been experimenting with carbon filaments since 1860, had been developing lamps with such filaments in Newcastle, and the question of priority in the use of this, the only really practical material at the time, was warmly debated. Swan's lamps were certainly well established in England before those of Edison, but litigation between the two parties concerned was wisely avoided by an agreement to amalgamate their interests. To this end the Edison and Swan United Electric Light Co. was registered on 26 October 1883, with a capital of £1,000,000, to acquire the British business and properties of the Edison Co. and the Swan Co. The claims of Swan were stated by the inventor himself in his paper on "Electric Lighting by Incandescence" read at the York meeting of the British Association in 1881, in the course of which he said:

This simple form of lamp I showed lighted at a lecture which I delivered before the Philosophical Society of Newcastle in February 1879. Very soon after this, and I am quite sure, without knowing what I was doing, Mr Edison produced a lamp identical with mine in all essential particulars. It, too, consisted of a simple bulb from which the air had been exhausted by the Sprengel pump and which, like mine, had no screw-closed openings nor complications of any kind, but contained simply the ingoing and outgoing wires sealed into the glass with the carbon attached to them.

The first public demonstration of electric lighting on a large scale by means of incandescent lamps was given in Newcastle on 10 October 1880, and during the next year many installations of Swan lamps were put into service, one of the most noteworthy being for the lighting of the Savoy Theatre in London, which was inaugurated on 28 December 1881. This constituted the first example of a theatre being entirely lit by electricity. The power plant consisted of a pair of Fowler engines arranged to drive six Siemens A.C. generators which supplied current for the 1,200 lamps.

A further striking proof of the fallacy of Professor Thompson's prediction was soon to be given. On 2 January 1882, Mr E. H. Johnson, Edison's agent in London, obtained the permission of the City Authorities to undertake the lighting of Holborn Viaduct and the neighbouring thoroughfares by means of Edison incandescent lamps. Once again, therefore, the Viaduct was destined to play a prominent part in the history of electric lighting. At the time in question the original gas lighting had been restored, and the object of the Edison enterprise was to demonstrate the advantages of the then novel system of incandescent lighting by replacing the gas burners by electric lamps mounted on the existing gas standards. The agreement with the Edison Co. provided that the Viaduct and adjacent streets should be lighted free of cost to the City for a period of three months commencing on 1 February 1882, and although the City Corporation had no power to authorize a supply being given to private consumers in the vicinity, it was understood that they would raise no objections to such a proceeding. Work was started almost at once, and it is recorded that the plant was running by January 12 but for various reasons the opening ceremony did not take place until April 12, and the commencement of the three months' free trial period was deferred until April 24. At its conclusion the Edison Co. made an arrangement with the City Authorities to continue the lighting for a further period of six months, at the same price as gas lighting. This period was subsequently extended and service was continued until the station was shut down in 1886.

The power station was in a building (No. 57) on the north side of the street, a few doors east of the end of the Viaduct. The first generating unit consisted of an Edison dynamo directly driven through a flexible coupling by a Porter-Allen horizontal steam engine mounted on the same base-plate. The complete unit weighed 22 tons. Steam was supplied by a water-tube boiler constructed by Messrs Babcock and Wilcox, who thus commenced their long connection with the power station industry. The engine had two cylinders, each 11 in. diameter by 16 in. stroke and developed 125 I.H.P. at 350 R.P.M. The dynamo was of the "Jumbo" type, with 12-limbed shunt-wound field magnets arranged horizontally and terminating in cast-iron pole-pieces between which the armature ran. The latter was 26·5 in. in diameter, and provided with a commutator 12·75 in. in diameter built up of 106 segments. Current was generated at 110 v., the voltage being regulated by a rheostat in series with the field. The dynamo was capable of serving 1,000 Edison lamps of 16 c.p. A second generating unit was soon added, this being of the same type but capable of serving 1,200 lamps. Later on, another 1,200-light set was put down at 35 Snow Hill, and a small 250-light set, driven by an Armington and Sims engine, was installed for the day load of the private consumers.

The public lighting extended from Holborn Circus to the Old Bailey and leads were soon carried to the General Post Office in Newgate Street, where some 400 incandescent lamps were installed in the Telegraph Operating Room. The system started with 938 lamps, mostly of 16 c.p. though some were of 8 c.p., illuminating the Viaduct and most of the premises on either side of the street. Amongst the buildings lighted was the famous City Temple of Dr Joseph Parker, situated at the west end of the Viaduct. This was equipped with 170 lamps, and has the credit of being the first church to be electrically lighted.

The mains consisted of a pair of segment-shaped copper conductors fixed in insulating material and carried in wrought-iron pipes. They were laid along the existing subway, and current for the street lamps and for private consumers was taken from them by insulated cables. Each circuit was protected by a fuse, and shunted electrolytic meters were used for determining the amount of current consumed.

The distinction of being the first Public Power Station in the world, if purely arc-lighting plants are disregarded, has been claimed for the Holborn Viaduct Station, for its current was supplied, not only for street lighting but to private consumers as well. Although established by Edison and his associates, it has an easy priority over anything of the kind in America, for the first Edison Station in the United States, namely, that of Pearl Street in New York, which was equipped similarly to the Holborn Station, namely with Babcock and Wilcox boilers, Porter-Allen engines and Edison "Jumbo" dynamos, was not inaugurated until 4 September 1882. A better claim, however, can be made for Godalming, in Surrey, where there was certainly a little Central Power Station working in 1881. It was installed and operated by Messrs Siemens Bros. and Co., under a yearly contract with the Town Council for the lighting of the streets, but current was also available for such private consumers as desired it. Power was developed from a waterfall on the river Wey, and the cables were laid in the gutters along the streets which were lit by both arc and incandescent lamps. The undertaking was an unprofitable one for the Company, and as a canvass of the citizens showed no prospect of a satisfactory demand, the supply was discontinued on 1 May 1884, and gas lighting restored to the streets.

If we confine ourselves to undertakings that have had a continuous existence to the present time, there seems no doubt that the honour of priority should be given to that of Brighton, which was offering to supply current to all consumers who desired it as early as February 1882. The history of the Brighton undertaking may be traced back to December 1881, when Mr Robert Hammond, who was so prominent a figure in electrical matters during the early days of the industry, came to Brighton to stage an exhibition of the Brush arc-lighting system in the town. This created so much interest that he agreed to leave the plant there for a time, in order that a circuit could be run along some of the principal streets of the town with the object of giving shopkeepers an opportunity of testing the advantages of the new light. This experimental circuit extended for $1\frac{3}{4}$ miles and carried 16 arc lamps in series, supplied with current from a Brush arc-lighting dynamo with an output of 10·5 amp. at 800 v., driven at 900 R.P.M. by a Robey engine of 12 N.H.P. The plant was erected in the yard of Reed's Iron Foundry, in Gloucester Road, and went into service on 21 January 1882. The demonstration was continued for a week, during which time applications were invited from people willing to pay 12s. per lamp per week for the illumination of their premises. The response appeared good enough to warrant the Hammond Electric Light Co. giving a permanent supply, which was inaugurated on 27 February 1882, and gradually the whole of the 16 lamps were contracted for. Current was available from dusk until 11 p.m. daily. By the spring of 1883 the demand had increased sufficiently to justify the installation of another arc-lighter, this time capable of serving a circuit of 40 lamps, driven by a Marshall engine. The Company then altered their method of charging and introduced the system of requiring payment of 6s. per lamp-week, plus an additional 1s. 6d. for every carbon consumed. This was probably the first instance of charging for electricity by a two-part tariff.

In the same year, Mr Arthur Wright, one of the most capable station engineers that the industry has produced, who was in charge of the Brighton undertaking from its inception until 1905, devised a scheme of running incandescent lamps from his arc circuits, and promptly put it into effect. Since the essential feature of these circuits was that they carried a constant current of 10·5 amp., which was far too much for any ordinary incandescent lamp, Mr Wright arranged groups of ten lamps, each taking about 1 amp., in parallel, and connected each group in series with the mains. The drawback to such an arrangement is, of course, that in case of the failure of one lamp, the voltage would rise across the remaining lamps of the group, with the almost certain result that they would rapidly fail one after the other. To meet this difficulty he at first connected a small electro-magnet in series with each lamp, so that if the lamp failed the cessation of current would allow an armature to drop into contact with a bar and thus switch a spare lamp into circuit, only one spare lamp being required for the whole group. The next improvement was to put an electro-magnet in the holder of each lamp. The fall of its armature made contact with a third wire leading to the spare lamp. This arrangement reduced the number of wires from eleven to three for a group of ten lamps in parallel. In its final form the protective device consisted of a single electro-magnet connected in parallel with the group of lamps. Its armature was delicately adjusted so that the rise of voltage due to the failure of a lamp would cause it to close a contact and throw into circuit a spare lamp of only half the resistance of the ordinary lamps. In the event of a further failure causing the voltage again to increase, the armature moved a little farther and closed a second contact which short-circuited all the lamps of the group. The lamps were not always in groups of ten, a smaller number of proportionately greater power being sometimes used in parallel, but the disadvantage of not being able to turn out individual lamps very much restricted the system as regards domestic lighting. Nevertheless, by January 1886 there were 1000 incandescent lamps on the 8 miles of arc circuits, and Brighton was the only town employing such a combination. Before the close of the next year there were 34 arc lamps and 1,500 incandescent lamps in service, all supplied with current from five Brush arc-lighting machines by overhead mains of no. 7 bare copper wire. The mains had a length of 15 miles, and extended to a distance of 3 miles from the Station. Indeed, prior to the introduction of transformers, Brighton furnished the most successful instance of long-distance transmission in any country.

We are now so accustomed to regard the voltage as the thing that has to be maintained constant in power station work, that it is not very easy to adjust one's mind to conditions of operation in which it is the current that has to be kept at some definite value and the voltage regulated accordingly. This, however, was what had to be done at Brighton, in common with other arc-lighting plants on the series system. At first the control of the voltage was entrusted to a boy who operated a carbon resistance connected as a shunt across the field of the dynamo. This was satisfactory enough for arc lighting, but it was found that the adjustment was neither prompt enough nor delicate enough when incandescent lamps were connected in the arc circuits. Moreover, the boy had to be on duty during the hours when he ought to have been in bed, and it is not to be wondered at if he occasionally fell asleep over his job. As an improvement on the carbon resistances, liquid rheostats were introduced at the suggestion of Mr James Swinburne in October 1884, but it was considered by Mr Wright that some improvement on the boy was also needed. He therefore devised an automatic regulator, the essential feature of which was a pair of movable electrodes which were raised or lowered in a tank of water by the action of a long solenoid energized by the current in the circuit. If the current decreased the electrodes were lifted, thereby increasing the resistance of the shunt, which they constituted across the field of the dynamo, and so bringing about a rise in the voltage of the machine, with a consequent restoration of the current to its normal value. As the electrodes approached either end of their travel a contact was closed, causing a bell to ring, when the attendant raised or lowered the speed of the engine as might be required. The bell circuits were so arranged that, when two or more dynamos were operating in series, any of their regulators could give the signal for an increase of engine speed, though no signal for a reduction of speed could be given unless all the regulators simultaneously permitted it. The automatic system of regulation was sensitive enough to keep the current within 1 % of its proper value under any conditions of load. The current was indicated by an ammeter consisting of a solenoid which pulled down a steel plunger against the action of a counterweight, and so caused the pointer to move across the scale. The latter had its central point calibrated to a current of 10·5 amp. in the solenoid, and the pointer reached the limit of the scale on either side of the centre if the current differed from the proper value by as little as plus or minus one-tenth of an ampere.

Mr Wright constructed other ammeters for use in the Brighton Station. In one very successful design a coil of wire traversed by the current to be measured, contained two iron cores, of which one was fixed while the other was free to rotate about its axis. The adjacent ends of the cores were bent over at right angles to form heads, which being magnetized to the same polarity by the current in the coil, repelled each other. The free core therefore was caused to rotate, and a pointer attached to it served to indicate the strength of the current. With such an instrument it was possible to get a scale of 140°, but it was found that the hysteresis of the iron cores caused the ammeter to read differently on a rising and falling load. This, however, was got over by short-circuiting the instrument between any two readings.

Home-made recording voltmeters were another outcome of Mr Wright's ingenuity, and of the requirements of the Station. These were constructed by using the works of a cheap alarm clock to pull along a strip of carbon-coated paper, the free end of which was kept taut by a weight. A solenoid in series with a resistance of 10,000 ohms was connected across the mains, and its plunger operated a spring-controlled needle which marked the paper. There were four such recorders in use, with charts to a scale of 100 v. per cm. and a travel of 2 cm. per hour.

House meters for measuring the current supplied to consumers were introduced by Mr Wright in the year 1884. His meters were of the electrolytic type, as were those of Edison in the Holborn Viaduct installation, but the constant current system of supply employed at Brighton made the problem of metering somewhat different in its nature. Edison had to assume the voltage constant, and to determine the number of ampere-hours consumed, whereas Wright could assume the current to be constant and measure the number of volt-hours. This was a considerable advantage, especially as regards accuracy, for it permitted the meter to be connected as a shunt across the circuit with a high resistance in series, and avoided the errors that might be incurred by the shunting of the meter itself, as was necessary in the Edison system. The meter consisted of a small glass jar containing a solution of sulphate of copper in which dipped two copper electrodes. The cathode was suspended from a counterweighted arm to which a pointer was attached. As the weight of the cathode increased, owing to the deposit of copper when the meter was working, the pointer moved over a scale and thus gave an indication to the consumer of the amount of current used since the last reading. Current was charged for at the rate of 1s. per K.W.H., the bill being based on the weight of copper deposited on

PLATE 1. BRIGHTON POWER STATION, 1887

From contemporary engraving in *The Electrician*

the cathode. The meter was connected in series with a resistance of several hundred ohms, composed of bare galvanized iron wire stretched over a wooden frame.

In 1885 the Hammond Electric Light and Power Co., who owned the undertaking, went into voluntary liquidation, and their Brighton property and interests were purchased by a new Company, the Brighton Electric Light Co., Ltd. This Company, which was registered on 16 December 1885 with a capital of £25,000, inherited from its predecessor a completely new Power Station adjoining the old one in the Gloucester Road, almost in the centre of the town. This Station was got into operation before the end of the year. The building was of brick, 70 ft. long by 55 ft. wide. The station started with three Brush arc-lighting dynamos, each designed to serve 40 lamps in series, a current of 10·5 amp. at 1,800 v. being required for this duty. The three machines were driven by belts from a countershaft, to which was belted a compound Fowler engine of the semi-portable type with cylinders underneath the front end of a locomotive boiler. This engine was capable of developing 200 I.H.P. and was operated daily from dusk until 1.0 a.m. Shortly afterwards two more Brush arc-lighting dynamos were added and another Fowler engine of the same type was installed. In 1887 it was decided to give a continuous service of electricity, and a 35 I.H.P. semi-portable Hornsby engine and a 16-light dynamo, capable of 10·5 amp. at 800 v., was put down for the day load.

The three engines all had the same maximum speed of 90 R.P.M. and all drove the same countershaft, which was placed at floor-level with the engines on one side and the five dynamos on the other. Like the engines, the dynamos all had the same maximum speed, which was 900 R.P.M. The principle of operation was to run the machinery always at the minimum speed possible for the output, this being thought to result in the most economical working. The daily routine of the station was as follows. During the hours of daylight the small Hornsby engine and the 16-light dynamo were sufficient to carry the load, all the circuits being connected in series. Towards evening one of the Fowler engines was started and the circuit divided into two parts, the larger part being put on one of the 40-light machines. As more lamps were switched on the voltage of this machine was raised by the action of the automatic shunt regulator. When the regulator reached its limiting position it caused a bell to ring. The attendant did not then increase the speed of the engine, but again divided the circuit and switched in a third dynamo. This process was continued until all the five dynamos were at work, each on its own separate circuit. After this, increases of load had to be met by increasing the engine speed. The engine, up to this time, had been running at only about 40 R.P.M., and its speed was raised, one revolution at a time, until the peak load was reached. As the load fell off, the operations were reversed, until at last the Hornsby engine was left alone to supply the town by a single circuit.

To bring a new dynamo into circuit it was first run up to speed with its armature shunted by a water rheostat and its field regulator set for minimum volts. The machine was then connected to two terminals on the switchboard through which the current was passing, and the short-circuiting plug between the terminals was withdrawn. This left the incoming dynamo in series with the circuit, though it was generating very little voltage. The shunt across its armature was next withdrawn until full current was passing through the armature, when the field regulator was allowed to come into action, causing the voltage of the machine to rise. The machine would then be working in series with another machine on a combined circuit. The separation of this into two independent circuits, each operated by its own dynamo, was done by simply turning a disconnecting switch after the voltage of each machine had been adjusted until it was just sufficient to drive the current through its own circuit.

All the engines worked non-condensing, their exhausts being turned into the chimneys of their boilers. The latter were fed with water from the town mains, whose initial hardness of 15° was reduced to 2° by means of a Stanhope softener, so that they only had to be washed out at intervals of three months.

The Company soon began to realize the limitations of the direct-current series-parallel system of working, and in order to cope with their growing business, they followed the example of Eastbourne and changed over to the high-tension alternating current system in 1887. This development was accompanied by the change of their name to the Brighton and Hove Electric Light Co., which took place in 1888. The new alternating current supply was generated at 1,800 v. by single-phase Lowrie-Hall machines, its pressure being reduced to 100 v. for the consumers by means of transformers erected on the house tops or placed in cellars or in street boxes as might be most convenient.

Ever since the commencement of the undertaking in 1881 it had been carried on without statutory authority, this being unnecessary so long as transmission and distribution were effected by overhead lines. The Brighton Corporation had themselves obtained a Provisional Order in 1883 for the supply of electricity in the same area, but had taken no steps to provide it. The position was, therefore, that the Company were working without authority while the Corporation were holding the authority without working. From 1884 onwards the Company made effort after effort to induce the Corporation to transfer the Order to them, so that they might replace their overhead lines by underground cables and also conduct their business in a more regular fashion, but had been uniformly unsuccessful. Finally in 1889, in view of what appeared to be the "dog in the manger" policy of the municipal authorities, the Company applied to the Board of Trade for a Provisional Order for themselves. The Corporation opposed the granting of such an order, with the result that a Board of Trade enquiry was held by Major Marindin. He declined the application of the Company, but made it clear to the Corporation that, unless they took steps to act upon the Order they were holding, it would be revoked. They therefore decided to establish a municipal service of electricity for Brighton, and in 1890 they set about the erection of a Power Station in North Road, practically opposite Reed's Iron Foundry. Faced with the threat of competition, the Company made the bold move of notifying their customers that their supply would be cut off unless they signed a contract to take current from the Company for a period of three years. This action, of course, was possible because the Company were not working under statutory powers. It is said that three-quarters of the consumers made the desired agreement. The municipal station was officially opened on 14 September 1891, and worked in competition with that of the Company until the rivalry was brought to an end by the Corporation arranging to buy out the Company in January 1894, on the understanding that it should continue operating until the end of the following September, during which time its customers would be transferred to the mains of the Corporation.

The first municipal supply was given by direct current at 115 v. on the two-wire system. The North Road Station started with four Willans-Goolden generating sets, two of them with a capacity of 45 k.w. at 450 r.p.m., and the other pair with a capacity of 120 k.w. at 350 r.p.m., all the engines being of the two-crank compound type. A battery of e.p.s. cells was provided to maintain the service while the machinery was shut down during the early hours of the morning. In the boiler-house were three Lancashire boilers, 7 ft. diameter by 28 ft. long, working at 150 lb. pressure. They had no economizers and the station worked non-condensing, the exhaust from the engines passing through a feed-heater and then being discharged up the chimney.

In 1893 the plant was extended and the supply changed over to the three-wire system at 115/230 v. Three more Lancashire boilers were installed, together with a pair of three-crank compound Willans engines driving Latimer Clark dynamos of 240 k.w. capacity. The latter were wound to generate at 230 v. and the earlier 120 k.w. sets were rewound for the same voltage, the smaller pair of units being operated as balancers across the two sides of the system. By 1904 the capacity of the plant had attained 5,935 k.w., when it comprised nineteen Willans engines directly coupled to continuous current generators of various sizes, and supplied with steam from six Lancashire and ten Babcock and Wilcox boilers. In 1908 the North Road Station was shut down, being superseded by a new municipal station at Southwick which was opened in June 1906. This was equipped on modern lines with turbo-alternators generating three-phase current at 8,000 v. and 50 cycles.

CHAPTER II

THE GROSVENOR GALLERY AND DEPTFORD STATIONS

"...Your young men shall see visions."

Joel, ii. 28

I F one station more than another may be considered as the real cradle of the modern power station industry, that honour must certainly be given to the station put down in the year 1883 by Sir Coutts Lindsay for the lighting of the Grosvenor Gallery in New Bond Street, London. This is not because the original plant, nor indeed anything about the station, was particularly remarkable, but rather on account of the extraordinary developments to which it gave rise under the genius of the late Dr Sebastian Ziani de Ferranti.

The first installation was of a temporary nature, consisting of a pair of Marshall's semi-portable engines erected in an outbuilding and belted to a couple of separately excited Siemens alternators generating single-phase current at 2,000 v. The plant was put down principally for the operation of arc lamps in series, and an automatic regulator kept the line current constant at 10 amp. It was not long before applications for a supply of electricity were received from certain shopkeepers and residents in the neighbourhood, and these were met by installing a transformer, or "secondary generator" as it was then called, in the house of each consumer. The primaries of all the transformers on a circuit were connected in series with the line, in accordance with the system then recently introduced by Gaulard and Gibbs, the high-tension current being transmitted by overhead cables supported from poles on the housetops.

In consequence of the growth of the demand from outside consumers, it was decided to establish a permanent generating station on a considerably larger scale, and the construction of this was commenced in December 1884. An excavation 65 ft. long by 51 ft. wide was made underneath the Gallery for the accommodation of the new machinery, and the boilers were housed in the basement of a new building connected with the engine room by a tunnel about 50 ft. long. Above the boiler room were offices and stores, and on the roof was a tank holding a three-days supply of feed water. The tunnel served as a ventilating duct for the engine room, air being withdrawn from it by a 72 in. Blackman fan running at 200 R.P.M. A chimney 110 ft. high with an internal cross-section of 36 sq. ft. was constructed for the discharge of the flue gases and exhaust steam. The chimney was designed for double the initial capacity of the plant.

The new station went into service towards the end of 1885. Its first equipment comprised a pair of Siemens single-phase alternators—the largest yet built by the Company—driven by belts from a countershaft to which any or all of three horizontal Marshall engines could be clutched, according to the load on the station. The two smaller engines were of similar size and design, each having a single cylinder of 19 in. diameter by 3 ft. stroke, and developing 225 I.H.P. at 80 R.P.M. The third and larger engine had two cylinders, each 19 in. diameter by 3 ft. stroke and was capable of developing 450 I.H.P. at 80 R.P.M. Current was sent out, as before, along overhead lines with house transformers in series for reducing the voltage to the consumers.

This arrangement gave a very great deal of trouble, and indeed proved practically unworkable. Ferranti, then a young man of only 21 years of age, who was at the time building alternators of his own design at a works in Appold Street, in the City of London, was called in to advise the Company. He entered the Company's service in January 1886, and took charge of the station, proceeding immediately to make radical alterations to the system. One of his first steps was to replace the Gaulard and Gibbs series transformers by others designed for working in parallel across the mains, thereby introducing the method which has since become universal. He also replaced the Siemens alternators by others of his own type wound with copper tape instead of round wire, and generating at 2,400 v. Their voltage was measured by the first electrostatic voltmeter ever constructed by Sir William Thomson or ever used in a power station. The current leaving the station was determined by a Thomson ampere balance working at the line voltage.

In August 1887 a new Company, called the London Electric Supply Corporation, Ltd., was formed to take over the Grosvenor Gallery Station from Sir Coutts Lindsay and Co., Ltd., as the nucleus of a scheme for supplying electricity to London on a really large scale. The Corporation had an authorized capital of £1,000,000 in shares of £5 each, of which £535,000 was subscribed. Almost the whole of this was put up by twenty-eight shareholders, of which the largest were Lord Wantage, V.C., and Sir Coutts Lindsay, who held shares to the value of £220,000 and £48,885 respectively, Lord Wantage thereby showing a financial courage comparable with the physical courage which had earned him the V.C. in the Crimean War. The project in view was the establishment of a vast generating station at Deptford, from which current for the lighting of 2,000,000 lamps would eventually be sent to London. Work on the Deptford site was begun in April 1888, but the story of this station must be told later.

Meanwhile, under Ferranti's energetic and skilful direction, the Grosvenor Gallery undertaking grew rapidly. In a short time there were five external circuits radiating from a tower on top of the Gallery. Each circuit was protected by fuses of stranded wire 2 ft. in length, and controlled by an air-break switch with an opening of 4 ft. The system increased until it covered an area extending from Regent's Park in the north to the Thames in the south, and from Knightsbridge in the west to the Law Courts in the east. The overhead cables were carried by hundreds of iron poles, $3\frac{3}{8}$ in. in diameter, fixed into cast-iron sockets on the tops of the houses. The return leads followed the same lines as the outgoing leads, being suspended about a foot below them. The larger cables were of 19/15 rubber-covered copper wire, suspended by leather thongs from cables of 7/16 steel wire. The service leads contained 7/16 and 7/20 conductors carried through earthenware tubes to the house transformers by way of fuses and double-pole switches. Nearly all the consumers' lamps were of 10 c.p., but there were a few arc lamps for commercial use.

Current was not metered to the consumers, the latter paying at the rate of £1 per annum for each 10 c.p. lamp connected, and at the rate of £2 per annum for each 20 c.p. lamp.

The station, in its reorganized state, started with two 2,400 v. single-phase Ferranti alternators, each designed for the supply of sufficient current for 10,000 lamps of 10 c.p. each. If such lamps had a consumption of 3·5 w. per c.p., which was the accepted figure in those days, the machines might be taken as having a rated output of about 400 K.W., allowing for the necessary losses. They were, however, very liberally proportioned, for it is recorded that one of them proved itself capable of maintaining no less than 19,500 10 c.p. lamps, which can hardly have been done with an output of less than 700 K.W. Whatever the maximum continuous rating of these alternators may have been according to modern methods of computation, they were enormous machines for the time. Each stood 9 ft. 6 in. in height and occupied a floor space of 9 by 11 ft. Including the base-plate and the 5 ft. driving pulley, the total weight of each alternator was 33·5 tons, of which 17 tons was accounted for by the field magnets, and 1·5 tons by the revolving armature which was 8·5 ft. in diameter. The machines had 40 magnet poles on each side and ran at 250 R.P.M. The fields were excited by shunt-wound Siemens dynamos coupled direct to the armature shafts. A 12 K.W. Kapp dynamo, which had served for lighting the restaurant of the Inventions Exhibition in 1885, was purchased by Sir Coutts Lindsay and installed for lighting the engine room or supplying current for excitation in case of need. The Ferranti alternators produced current at a frequency of $83\frac{1}{3}$ cycles per second. The choice of such an odd frequency may appear curious at first sight, but it must be remembered that this figure corresponds to 10,000 reversals of the current per minute, which was the way in which frequency was considered in the earliest days of the industry.

One of the main alternators was driven by leather belting from the countershaft to which the three Marshall engines could be connected by clutches as already mentioned. The other was driven by ropes from the 18 ft. flywheel of a 700 I.H.P. Corliss engine built by Messrs Hick, Hargreaves and Co. This was a horizontal engine with a single cylinder 33 in. diameter by 4 ft. stroke. The method of operating the alternators was as follows. At times of light load the belt-driven machine was actuated by one of the smaller Marshall engines clutched to the countershaft. As this engine approached the limit of its capacity, the other small one was run up to speed and clutched to the shaft as well. This required considerable skill on the part of the operator, who had to bring the speed of the incoming engine to an exact equality with that of the shaft, and at the same time to get the claws of the clutch in the proper relative position so that he could slide the movable part into engagement with the other. When the desired moment approached, he brought the teeth of the two parts very gently into contact and was thus able to feel when engagement could be safely effected. When, in due course, this second engine also became fully loaded, the operator clutched in the 450 H.P. engine in the same manner. All the three engines were then driving one of the alternators, and, as soon as further power was required, the, 700 H.P. Corliss set was run up and made to take all the load it could. There was no attempt made to run the two alternators in parallel, even for the transfer of the load from one machine to another. The changing over of the load was carried out by the simple method of disconnecting the feeder circuits one by one from the first alternator and connecting them to the second. The switches were arranged so that this could be done without causing more than a momentary blink in the lights. After the transfer of the load to the Corliss unit, one or more of the Marshall engines could be disconnected from the countershaft and shut down, only to be restarted in the event of a further increase of load, or when the falling off in the demand for power permitted all the load to be transferred back to the Marshall engines and the large Corliss set shut down. The object of operating in this way was, of course, to have every engine in service as far as possible fully loaded and thus to secure the highest economy. The amount of load taken by each of the engines driving the countershaft was regulated by moving a weight along the arm of the governor lever, and the voltage of each alternator was indicated by a Cardew hot-wire voltmeter. Lubrication of the machines was effected by castor oil flowing to the bearings by gravity from a 300-gallon tank over-head, to which it was returned, after being strained, by means of a pair of Worthington pumps.

Steam was supplied to the engines from a battery of four Babcock and Wilcox water-tube boilers which worked at a pressure of 130 lb. per sq. in. Each boiler had a rated capacity of 150 boiler H.P., and was therefore deemed capable of delivering about 5,000 lb. of steam per hour. The whole of the boiler power was needed at times of full load, and steaming was almost continuous, the fires being drawn only for a few hours on Sundays. The average daily consumption of coal attained a figure of about 23 tons, and room only existed for a storage of 36 tons.

The Company had been sanguine enough to believe that their great new station at Deptford would have been in a position to deliver current by the end of 1888, and in accordance with this anticipation they had taken on customers for the extra supply that was foreseen. The hopes were badly falsified, with the result that the new customers had to be served by the Grosvenor Gallery Station, which soon became badly overloaded. No current was, in fact, transmitted between Deptford and the Grosvenor Gallery before October 1889, and then it was sent from the latter station to Deptford for the lighting of the works. It was stepped up from 2,400 to 5,000 v. for transmission, and then stepped down to 100 v. by two stages after arriving at its destination. The next year was spent mainly in trying to overcome troubles with the 10,000 v. cables and with the new machinery at Deptford, until, on 15 November 1890, a serious disaster overtook the Grosvenor Gallery Station. The engine-room machinery which had been shut down for good on November 1 was being dismantled at the time, as the station had been converted into a transformer substation, to deal with current coming from Deptford at high voltage. A bank of transformers had been erected temporarily in the only available place, a room with wood-lined walls and ceiling over the boilers and beneath the water tank. Early in the morning of the fatal day a man, in attempting to bring a fresh set of transformers into service, started an arc at a plug switch, and then, unthinkingly, withdrew the plug. The consequence was the formation of a fierce 5000 v. arc which set on fire the temporary india-rubber covered cables, and then everything else, so that in half an hour the whole place, together with its contents, was completely destroyed. The catastrophe was, in a sense, a needless one, for had the man kept his head he could have interrupted the current immediately by a circuit-breaker, or cut off the supply from Deptford at the main switch.

Another set of transformers, some new and some which had been hastily repaired after the fire, was assembled in the old engine room with fire-proof precautions and supply was resumed on nearly all circuits by November 26. On December 3, however, one of the repaired transformers failed, and transferred its load to the remainder which, being fully loaded at the time, promptly burnt out, one after the other. In view of this second disaster, and because Deptford was really not in a condition to carry the load, it was decided that the only thing to do was to cease all attempts to give a supply until a reliable one could be assured. The Company was serving 312 customers at the time, and had a total of 38,272 lamps connected to its system, of which number more than 20,000 were sometimes in use simultaneously.

The Deptford Station, mentioned above, will always be associated with the name of Ferranti. Before he had been very long at the Grosvenor Gallery he had visions of supplying the whole of London from one great power station which should be situated at some place where land was cheap, where unlimited water was available, where sea-borne coal could be obtained at low prices, and where operations could be conducted on a scale quite impossible in the midst of a residential or business area. The idea came to him while studying a map of London, when he noticed the tendency of the gas undertakings to shut down their small isolated works in the better class districts and concentrate their plant at such places as Beckton and Rotherhithe. In fact, in a report made on 31 May 1890, in justification of his scheme, he said: "It may be well to remind the shareholders that, in erecting its works at Deptford on a site immediately adjacent to the Thames, and upon a sufficiently comprehensive scale, the Corporation has

taken its lesson from the Gas Companies." It was at Deptford
that he found a site offering the facilities he sought, and he decided that the proper plan would be to generate electricity there
in bulk and transmit it to substations in London by high-voltage
cables. The scheme involved transmission of current at the then
unheard-of pressure of 10,000 v., but Ferranti could see no in-
superable difficulty in doing this, and his directors were so convinced by his confidence in his plans and by his faith in his ability
to carry them to a successful conclusion that on 26 August 1887,
the London Electric Supply Corporation was registered with
a capital of £1,000,000 to take over the Grosvenor Gallery
Station and to put the larger project into effect.

If the prescience of Ferranti as to the form that electrical development would eventually take was little short of marvellous,
his engineering courage in attempting to realize his vision was no
less wonderful. He designed a station at Deptford with an ultimate capacity of 120,000 H.P., with units of 10,000 H.P. generating at 10,000 v. To assure himself as to the practicability of such
a voltage, when preparing his plans for the Deptford Station in
1888 he made a transformer to step up the Grosvenor Gallery
pressure of 2,400 to 10,000 v. for experimental purposes. At
that time there was no practical means of measuring such a
voltage, so Ferranti determined it with sufficient accuracy by
connecting a hundred 100 v. lamps in series across the high
tension terminals of the transformer and noting their brightness.
A demonstration of this kind was given before the Commission
appointed to report on the possibility of developing power from
the Falls of Niagara, with the object of proving the existence of
the voltage in question.

The Deptford Station, which was designed entirely by Ferranti,
was conceived on a scale immeasurably greater than anything of
the kind then existing in this or any other country. The main
building, 210 ft. long by 195 ft. wide and 100 ft. high, was
erected on a concrete raft 4 ft. thick at a distance of about 350 ft.
from the river. A quay 195 ft. long, served by a 50-ton crane, was
constructed on the river bank and connected to the station by a
railway which ran through the centre of the building, to give

facilities for getting in the machinery. There were two boiler-houses, one above the other, each with its own basement for the accommodation of the flues, ash-runways and fans for the forced draught, while immediately under the roof was a series of hoppers with sloping floors, providing storage capacity for 40,000 tons of coal. Runways to the quay enabled coal to be brought directly from the ships to the hoppers. Four flues connected the boilers to two rectangular chimneys 28 by 16 ft. in external cross-section and 150 ft. high, one at each end of the building. Each boiler-house was designed to contain two rows of Babcock and Wilcox hand-fired land-type boilers fronting on to a central firing aisle. Altogether there were to have been 80 boilers, each with a heating surface of 5,000 sq. ft. apart from the Green's economizers with which they were provided, the complete installation having a rated evaporative capacity of 1,380,000 lb. of steam per hour. It was intended, later, to triplicate the above arrangement. The boilers were designed for a working pressure of 200 lb. per sq. in., which was greatly in advance of current practice in those days.

The boiler plant was, of course, nearest the river, and on the other side of the longitudinal dividing wall running through the station were two engine-rooms parallel to each other, each 66 ft. in width and separated by a row of columns 80 ft. high. Beneath the engine-rooms was a basement 12 ft. deep. Half of the first engine-room, served by a 25-ton travelling crane, was to be occupied by a pair of 1,250 H.P. Corliss engines with 21 ft. pulleys running at 80 R.P.M., each driving a 5,000 v. Ferranti alternator at 120 R.P.M. by means of 40 cotton ropes. These units were to provide the first supply from the station, and afterwards to carry the comparatively light day load. The other half of the same engine-room was to house the condensing plant for the whole station. In the second engine-room, which was served by a 50-ton crane, were to be placed four 10,000 H.P. engines each directly coupled to a 10,000 v. Ferranti alternator.

In order to avoid the protracted negotiations which would have been necessary to obtain the consent of the numerous local authorities concerned for carrying underground cables through their streets from Deptford to London, even supposing that this

permission could have been obtained, it was decided to approach
the various Railway Companies serving the district, with a view
to carrying the cables along their lines, which were, for practical
purposes, their private property. The Railway Companies made
no objections, and agreements were come to with the South-
Eastern, the London, Brighton and South Coast, and the London,
Chatham and Dover Companies by which the mains could be
brought to London on their rights of way, and taken across the
bridges to Charing Cross, Cannon Street and Blackfriars Stations.
The Metropolitan and District Underground Railways likewise
gave permission for the use of their tunnels, so that access was
obtained to the principal centres of distribution. Thereafter, it was
proposed to use overhead lines, until Provisional Orders could be
obtained from the Board of Trade for laying cables in the streets.
Powers were applied for to use the streets belonging to 24
different Local Authorities, and this wholesale application, simultaneously with applications from other Companies, resulted in
the Board of Trade holding a local enquiry in April 1889 which
lasted no less than 18 days. As the outcome of this, the London
Electric Supply Corporation were granted Provisional Orders
concerning more than half the areas which they sought to supply.
They could not, however, be assured of a monopoly of supply in
any areas, for Major Marindin, who conducted the inquiry, held
the opinion that, wherever possible, two different Companies,
one supplying direct current and the other supplying alternating
current, should be allowed to compete in the same district. It is
somewhat surprising that this view was also held by the Chairman
of the London Electric Supply Corporation, who stated at a
meeting of the shareholders that the Directors were opposed in
principle to a monopoly for themselves or for anybody else. A
further result of the enquiry was the decision of the Board of
Trade that it would be unsafe for the London Electric Supply
Corporation to depend entirely upon one station so far away as
Deptford, and in consequence of this view the Directors agreed to
erect only two of the 10,000 H.P. units at Deptford, and to build
another station somewhere else for the other two.

Such, in outline, were the plans for the Deptford undertaking,

PLATE II. DEPTFORD POWER STATION, 1889

From contemporary engraving in *The Engineer*

and the circumstances under which a supply was to have been given. Work on the site was commenced in April 1888, and carried on night and day. By midsummer of the next year the wharf was finished and equipped, the main building was up and 24 Babcock and Wilcox boilers with an aggregate rated evaporative capacity of 414,000 lb. of steam per hour had been installed. The two 1,250 H.P. engines had been erected and run, and the alternators for them were almost ready. These engines, which had been constructed by Messrs Hick, Hargreaves and Co., Ltd., of Bolton, were of the vertical Corliss type with cylinders 28 in. and 56 in. in diameter by 4 ft. stroke. Each engine had a flywheel 22 ft. in diameter weighing 60 tons between the two cylinders, from which its alternator was driven by means of 40 cotton ropes 5 in. in circumference. The alternators, which were wound for 5,000 v. stood 14 ft. 6 in. high from their base-plates and had armatures 13 ft. in diameter. The main bearings of both engine and alternator were spherically seated and cooled by the circulation of water through them. They were lubricated under a gravity head from an elevated tank to which the oil was returned by pumps driven from the end of the alternator shaft. The alternators were excited at 50 v. by Kapp dynamos of 20 K.W. driven directly at 200 R.P.M. by Allen engines.

The two 10,000 H.P. engines and their 10,000 v. alternators were also in an advanced stage of construction. The engines, which were being built by Messrs Hick, Hargreaves and Co., Ltd., were of the vertical twin-tandem compound design, with the armature of the alternator arranged as a flywheel on the centre of the shaft. Each engine had 2 H.P. cylinders of 44 in. diameter and 2 L.P. cylinders of 88 in. diameter, all with a stroke of 6 ft. 3 in. The speed was 60 R.P.M. The engines were fitted with Corliss valve gear and were designed to work with steam at 200 lb. pressure. Their crankshafts were forged from ingots of steel weighing 75 tons each, these being the heaviest ingots cast up to that time in Great Britain. The shafts were bored from end to end with a central hole 12 in. in diameter. Each had a finished weight of 20 tons and a finished diameter of 36 in. at the centre. They ran in spherically seated bearings, lubricated under a gravity head.

The dimensions of the alternators were as imposing as those of the engines. They weighed 500 tons apiece, of which 225 tons was accounted for by the armature and shaft. The armatures were 46 ft. in diameter over the coils which were mounted on a ring 35 ft. in diameter. As a measure of safety for the operators, the collecting brushes of the alternators were to have been enclosed in iron boxes which were magnetically locked by the exciting current of the machines.

These immense generators were designed by Ferranti, and except for the castings required, the whole of their construction was carried out at Deptford under his supervision. Even the turning of the great shafts was done at the power station. Among the heavy machine tools installed for the manufacture of the alternators was a planer capable of dealing with work 20 ft. high by 22 ft. long, and a lathe able to turn pieces up to 11 ft. diameter by 25 ft. long. Owing to circumstances that will be made clear later, none of these great alternators was ever completed, so that one can only speculate as to what their performance would have been. The same has to be said about the engines built to drive them, but their story stands, nevertheless, as evidence of one of the boldest conceptions in the whole history of central station work. It may be well that they were fated to remain unfinished, for even when they were being designed there was already, for the few who could then read it, the "writing on the wall" foretelling the disappearance of the large slow-speed unit in favour of the type of generating machinery at that time being evolved by Parsons on the Tyneside.

Returning now to the 1,250 H.P. sets that were actually put in operation, to utilize their current, mains had been purchased and laid in the early part of 1889, but these had proved an almost complete failure. The consequence was that, when the units in question, together with their necessary boilers and auxiliaries, were ready to run, there was no reliable means of getting their current to London. Ferranti came to the conclusion that the only way to get satisfactory mains for the transmission of current at 10,000 v. was to make them himself in accordance with his own ideas. His design consisted of two concentric brazed copper

tubes of equal cross-sectional area, separated by paper im-
pregnated with ozokerite wax. The mains most generally used
had tubes with a cross-sectional area of 0·25 sq. in. and were
rated to carry 250 amp. They were made in 20 ft. lengths in
the following manner. The inner tube, after being cleaned
and straightened, was mounted in a simple machine which
rotated it, while bands of brown paper saturated with melted
wax were wound round it, to a thickness of half an inch. The
tube so prepared was next placed inside the outer conductor,
into which it fitted quite loosely, and the whole combination was
drawn through taper dies until the outer tube was in tight contact with the insulation. It was Ferranti's intention to leave the
outer conductor tube bare, and to keep it well earthed when laid,
but the Post Office authorities raised strong objections to this
course on the grounds of the interference that it would cause to
telephonic and telegraphic services. Ferranti, therefore, wound the
outer tube, which was $1\frac{21}{32}$ in. in diameter, with waxed paper to a
thickness of 0·125 in., and slipped the main thus insulated into a
thin iron tube of $2\frac{3}{8}$ in. external diameter with brazed seam. This
was then placed over an open fire, and melted wax was pumped
through a hole at the centre of its length. The wax flowed to
both ends and filled the space between the wrapping and the
iron tube surrounding it.

The paper insulation at one end of each 20 ft. length of main
was turned down to a male cone 6 in. long, and the other end
bored out to a hollow cone of the same length. The mains were
laid underground in wooden troughs in which they were kept
separate by wooden bearers. The troughs were afterwards filled
with pitch and closed by wooden covers. As the 20 ft. lengths of
main were laid in the troughs a copper sleeve 16 in. long and an
iron sleeve were slipped over one end of each length. A copper
plug 11 in. long was then pushed partly into the end of the inner
tube of the length to be connected up, and the male and female
coned ends were forced together by hydraulic pressure. The plug
made a perfect electrical connection for the inner conductor. The
copper sleeve was next brought over the joint and made to grip
the ends of the outer conductor by being corrugated on to them

by a special tool. This being done, the iron sleeve was put into position and its ends forced into contact with the ends of the iron casing tubes by means of a similar corrugating process. Melted wax was then forced in between the iron sleeve and the copper sleeve, and the joint was complete. To allow for expansion due to temperature changes, lengths of the main were laid at intervals with a slight double curvature, which rendered them sufficiently flexible. Altogether some 28 miles of this type of main were laid, and after the initial difficulties had been overcome, they worked satisfactorily for years at a pressure of 10,000 v., in spite of the thousands of joints involved in their construction. Some of the Ferranti mains, which were the pioneers of all the high-tension paper-insulated cables in the world, had a working life of more than 40 years. The electrical characteristics of these mains are worth recording. The resistance of the 0·25 in. main was 0·324 ohm per mile, including lead and return. The dielectric resistance between inner and outer conductor was 700 megohms per mile; the inductance was 286μH. per mile; the capacitance between the conductors was $0·367\mu$F. per mile, and the capacitance between the outer conductor and the iron casing was ten times as much.

The Board of Trade were at first much concerned about the safety of operating with cables at 10,000 v., while Ferranti insisted that perfect security was provided by maintaining the outer conductor at earth potential. To prove the truth of his contention the authorities were invited to witness a demonstration, which consisted in driving a cold chisel right through the main while it was alive at 10,000 v. Mr Harold W. Kolle, who had joined Ferranti in 1888 after being for three years the Electrical Engineer at Eastbourne, volunteered to hold the chisel while the blows were struck by another assistant, Mr C. Henty, who was the son of the famous war correspondent and writer of boys' stories, G. A. Henty. It was a courageous thing to do, but it demonstrated beyond question that confidence in the safety of the cables was justified, and the desired permission for them to be laid under roadways was therefore obtained. Mr Kolle, who later on became Chief Engineer and eventually a Director of Babcock and Wilcox Limited, is happily still alive to tell the tale of the historic test in which he participated some fifty years ago.

A curious phenomenon was observed when these cables were first put into service to carry current from Deptford to London. The voltage at the London end was found to be considerably greater than the voltage applied at the Deptford end. In fact, 8,500 v. between the conductors at Deptford was sufficient to produce 10,000 v. at London. This rise of voltage became known as the "Ferranti Effect". Something of the sort had been observed by Ferranti himself while the cables were being laid. They were then tested by a potential difference of 500 v. derived from a transformer at one end, and to be sure that the pressure was on the main a bank of five 100 v. lamps was connected in series across the other end. As the length of the main increased, it was noticed that the lamps burned brighter, and in time an extra lamp had to be placed in the bank to prevent the failure of the others. The "Ferranti Effect" only occurred when the mains were fed by transformers, the heaping up of the potential being due to the interaction between the self-induction of the transformers and the capacity of the mains.

Owing to one difficulty after another, no circuit could be established between the stations at Deptford and the Grosvenor Gallery before October 1889, and then it was only useful for sending current to Deptford for the lighting of the works. As already mentioned, serious trouble had developed with the original purchased mains, even at the 5,000 v. at which they were then operated, and Ferranti reported, on 31 May 1890, that until these mains could be replaced by the new type, it would not be possible to utilize more than a quarter of the generating capacity then available at Deptford. The two 10,000 H.P. engines and the boilers to serve them, had already been constructed, and as their 10,000 v. alternators were in a fairly advanced state, it was hoped to have these large additional sets ready for use in a fairly short time. Then, with the new mains in service, the full capacity of all this machinery would be at the disposal of the Company.

These hopes, however, were doomed to disappointment. The

station struggled on, working at times in parallel with that at the Grosvenor Gallery, until it was considered safe to dismantle the machinery at the latter station and convert the place into a sub-station for transforming and distributing the current from Dept-ford. This was the state of affairs on 15 November 1890, when the whole of the transforming apparatus at Grosvenor Gallery was burnt out, and that plant, therefore, put completely out of service. An abortive attempt was made to restart it as a sub-station, as already recounted, but the failure of this, together with other troubles in connection with the Deptford Station, which had not yet received delivery of either of its 10,000 H.P. engines, compelled the Company to stop the supply of electricity altogether for a period of three months.

Advantage was taken of this misfortune to remove the whole of the overhead transmission lines radiating from the Grosvenor Gallery and replace them by underground cables. Meanwhile the new 10,000 v. cables from Deptford were completed and the supply was restored over them on 15 February 1891, by means of the two 400 K.W. Ferranti alternators which had been trans-ferred to Deptford from the Grosvenor Gallery Station. They were driven by a pair of new Corliss engines of 700 I.H.P. put down for the purpose. These were compound engines with the H.P. and L.P. cylinders in tandem, and each drove its own alter-nator by 17 ropes from a 24 ft. flywheel weighing 35 tons. As the alternators were wound for the old Grosvenor Gallery pressure of 2,400 v., each was provided with a set of four trans-formers by which the pressure was raised to 10,000 v. for trans-mission. Switching was done at this voltage, the current passing first through an ammeter and then to the busbars by means of a hand-operated switch with an air-break of 36 in. There was an ammeter in each main, and protection was afforded by fuses which were at first mounted in long porcelain tubes, but later in tubes of glass in order that the fuse wires might be visible. Early in 1891 it was the custom to operate the alternators in parallel, to prevent a recurrence of the complaints which had sometimes been received from the customers of the Grosvenor Gallery Station because of the blinking of the lights at the times of changing over machines. There were, at first, only the two old alternators in service because the two 1,250 H.P. units were taken out of commission to enable their alternators to be rewound with the object of raising their voltage from 5,000 to 10,000 v. This work was not completed before August 1891, the whole of the load until then being carried by the two smaller machines. This was easily possible, for the three months' interruption of supply had caused many customers to transfer to other Companies, and when current was again available there were no more than 9,000 lamps remaining to be served, or less than a quarter of the number previously dependent on the system.

The misfortunes of the London Electric Supply Corporation were not without influence on the minds of the Directors. The Company had lost a large amount of money, and as the prospects of obtaining a virtual monopoly of the electricity supply of London began to fade, it is no wonder that the Board began to question the wisdom of Ferranti's policy. It must also be remembered that the whole principle of generating and transmitting current in bulk at so high a voltage and for so great a distance had met with almost universal criticism at the hands of men who were looked upon as the greatest authorities on the subject. A study of contemporary technical literature shows how few were those who regarded the scheme as otherwise than a reckless one. Many engineers, indeed, disbelieved in any future for alternating current supply at all, curious as this may seem at the present day. Even J. E. H. Gordon, deeply committed as he had been to alternating current practice in the early days, and well known as the designer of the famous machines at Paddington, had gone over to the ranks of the advocates of continuous current supply. Speaking at a technical meeting on 23 February 1888, he had stated that, after two years of operating experience with alternating current at Paddington, he had become so convinced of the superiority of supplying electricity by means of storage batteries that he was prepared to abandon his own system and every patent that he held. It may also be mentioned that Edison, whose engineering enterprise was above suspicion, expressed his disapproval of the scheme after paying a visit to

Deptford in October 1889. We know, now, that Ferranti's in-
stincts were right. In proof of this there stands to-day on the site
he chose at Deptford, a plant of more than 300,000 K.W. capacity,
the lineal descendant of the station that he founded in the eighties
of the last century. There is no doubt of the soundness of the
principles by which he was guided, but he was in advance of his
time, and Fate was against him. In May 1891 the Board gave
orders for all work to be suspended on the 10,000 H.P. engines
and generators. The engines, therefore, were never sent to Deptford, nor were the boilers for them ever installed, and in August
1891 Ferranti's connection with the Company came to an end.

After Ferranti's departure, his place as Chief Engineer to the
London Electric Supply Corporation was filled by Mr P. W.
d'Alton, who had been with the Nordenfeldt Submarine Company before joining Ferranti. Mr Gerald W. Partridge, who had
had charge of the electrical side at Deptford since 1888, soon
succeeded Mr d'Alton as Chief Engineer and later on became
Managing Director of the Company. He is still on the Board,
after having served the Company for more than fifty years.
Among other "old timers" of the London Company must be
mentioned Mr C. P. Sparks who, with Mr Kolle, was one of
Ferranti's assistants during the construction of the Deptford
Station and the laying of the 10,000 v. mains to the Grosvenor
Gallery Station. Mr Sparks remained with Ferranti until 1899, and
then, besides carrying on an extensive consulting practice, became
Chief Engineer to the County of London Electric Supply Co., a
position which he held for more than twenty years. Like Ferranti,
his eminence in the profession was acknowledged by his election
to the Presidency of the Institution of Electrical Engineers.

The departure of Ferranti took place at a critical time in the
life of the great undertaking that he had created. By the end of
August there were 30,000 lamps connected to the system, and
four transformer substations were in regular operation at the
Grosvenor Gallery, Trafalgar Square, Blackfriars and Deptford
respectively. The Deptford Generating Station was, however, far
from being ready for continuous service, even on the restricted
scale which then was alone possible, and the engines could only
run exhausting to the atmosphere because no condensing plant

had yet been installed. During the autumn numerous interruptions of the supply were caused by failures of either the alternators, the mains or the transformers, many of these failures being attributed to attempts to work in parallel. The transformer trouble was so serious that the Company found it necessary to establish their own transformer repair shop in the Deptford Station. While the staff were still struggling against continual misfortunes, a long-continued spell of fog and frost in November 1891 put the plant to a test which it was not in a position to withstand. Under the heavy and sustained load that it was called upon to carry, a series of breakdowns occurred, with the result that the system was once more completely shut down. In the words of Mr James Staats Forbes, the chairman of the Company, at a meeting of the shareholders: "The whole thing came to a collapse; the dynamos, mains and everything went wrong, and for four days or more we were without light." This was one of the worst of many blows of Fate, but another and a totally undeserved one fell in the following March when all the four Ferranti 10,000 v. mains between Deptford and London were put out of commission simultaneously. The cause was a fire which broke out in material stored beneath some arches of the South-Eastern Railway Co. near Spa Road, Bermondsey. The mains were carried along the parapet of the arches, but the flames were so intense that the exposed sections of the mains were completely destroyed. To guard against any similar accident in the future, the mains were subsequently placed underground throughout their entire length.

It was not until the end of February 1893 that the engines could be operated under condensing conditions, but the confidence of the public in the reliability of the supply had been so shaken by the repeated failures of current that there was not sufficient load for the large engines to be employed economically during many hours of the day. Consequently, a pair of 300 I.H.P. triple-expansion condensing units were installed, as no greater capacity than this was required to supply the early morning demand for a considerable portion of the year. It must have been a sad disappointment to have to put down sets of so small a size instead of erecting the 10,000 H.P. units which were to have been running before this time, but the claims of economy were urgent. In 1894 the Company found itself unable to repay a loan of £50,000 advanced two years before by Lord Wantage on the security of the undertaking. To protect his own interests, which included about £250,000 of ordinary and preference shares in the Company in addition to the £50,000 in question, and to safe-guard at the same time the interests of the other shareholders, Lord Wantage arranged for a receiver, nominated by himself, to take over the management of the business. Under the new re-gime important savings were effected, and a bold effort was made to rescue the undertaking from the fate which many people con-sidered inevitable. Much operating trouble at Deptford had been due to the foul nature of the Thames water used for boiler feeding, which was contaminated by salt and other impurities. To remedy this, a 6 in. bore-hole was sunk, to give a supply of pure water. In 1895 a new generating unit was ordered, con-sisting of a Ferranti alternator with a capacity of 1,000 K.W. at 11,000 V. driven at 156 R.P.M. by a three-crank compound vertical condensing engine. The engine was constructed by Messrs Plenty and Co. of Newbury, and was guaranteed to have a con-sumption of not more than 20 lb. of steam per I.H.P. when supplied with steam at 140 lb. pressure. The design was really an assembly of three tandem compound engines side by side on the same crankshaft. The H.P. and L.P. cylinders had diameters of 17·5 and 36 in. respectively, with a common stroke of 26 in. Steam was admitted to the H.P. cylinder by a piston valve, and to the L.P. cylinder by a D slide valve. Both valves were on the same spindle, and their weight was taken by a steam balance piston. The throw of the eccentric which operated each line of valves was controlled by Joy's hydraulic gear. The eccentric sheaf had an elongated hole through the centre, and was acted on by springs which tended to keep it central with the shaft, when, of course, no travel would be given to the valves. The action of the springs was opposed by a small hydraulic cylinder supplied with oil under pressure through a hole in the centre of the shaft. The pressure of this oil, and therefore the throw of the eccentric at any time, was under the control of the centrifugal governor of the engine.

The interior of each of the three cast-iron columns supporting the cylinders of the engine was used as a jet condenser for the steam exhausted from the cylinder above. Each condenser had its own air pump at the base of the column, operated by a lever from the cross-head. The cooling water was sucked into the condensers by vacuum from a large tank which was filled by the Thames at high tide, but the discharge was not returned to the tank except in very severe weather when it was necessary to prevent the tank freezing.

The alternator was driven directly from one end of the crankshaft of the engine. It was of the rotating armature type, in contrast to the machine recently constructed by Ferranti for the Portsmouth Power Station. The armature consisted of 64 flat coils mounted in pairs and supported by ebonite insulators from the rim of the flywheel. The latter was made in halves held together by shrunk hoops. The diameter over the coils was 22 ft., and the weight of the complete armature was 35 tons. It ran in spherically seated bearings, so beloved by Ferranti, which were 12 in. in diameter and 36 in. long. The field coils were carried on 5 in. cores of hammered iron, projecting horizontally inwards from the cast-iron side frames of the alternator.

Assisted by the various improvements brought about in the equipment and management of the undertaking, the corner was eventually turned, thanks to the indomitable tenacity of Lord Wantage and other large shareholders, who were likened by one of the American technical journals to men holding a tiger by the tail and wondering whether it would be safer to hang on or to let go. The gradual increase of the load, however, and the overcoming of the difficulties of high-tension generation and transmission, pointed to the success that was ultimately obtained. To put the Company on a sounder financial basis, the capital, which had been raised to £1,250,000 in 1889 by the issue of preference shares, was written down in 1898 to £850,000, thereby admitting that no less than £400,000 had been lost up to that time. The holders of the ordinary shares never received a penny on their investments until 1905, when the first dividend, amounting to 3 % on the reduced capital, was paid in respect of the previous year's working.

CHAPTER III

THE POWER STATION OF THE GREAT WESTERN RAILWAY CO. AT PADDINGTON

Twinkle, twinkle, little Arc,
Sickly, blue, uncertain spark;
Up above my head you swing,
Ugly, strange, expensive thing.

Now the flaring gas is gone
From the realms of Paddington,
You must show your quivering light,
Twinkle, blinkle, left and right.

Cold, unlovely, blinding star,
I've no notion what you are,
How your wondrous "system" works,
Who controls its jumps and jerks.

Though your light perchance surpass
Homely oil or vulgar gas,
Still, (I close with this remark)
I detest you, little Arc.

From "Lines to the Electric Light at the G.W. Railway Terminus" in the
St James' Gazette, 1888.

ONE of the active pioneers of the power station industry was Mr J. E. H. Gordon, who, after a brilliant career at Cambridge, entered the service of the Telegraph Construction and Maintenance Co. of Greenwich, and became the manager of their Electric Light Department. He was fortunate in possessing the confidence of Sir Daniel Gooch, who was Chairman both of that Company and also of the Great Western Railway Co., a combination of functions that was destined, as will be seen, to have important results. Towards the end of 1882, and a little prior to the appearance of the first Ferranti machine, Gordon constructed, to his own designs, what was then the largest electrical generator in the world, for the illumination of the Company's extensive cable factory and marine engine works at Enderby's Wharf, East Greenwich. It was an alternator, and said to be capable of supplying current to as many as 5,000 Swan lamps of 20 c.p., so that it may be presumed to have had a maximum capacity of something approaching 350 K.w. The machine was provided with a separate steam-driven exciter, and it was operated by a man in a small closed room who regulated the supply of steam to the engines in accordance with the indications of a photometer. All the instruments he had to assist him were a tachometer showing the speed of the alternator, an Ayrton ammeter to indicate the exciting current, and a steam pressure gauge. The photometer consisted of a screen on which two shadows of an iron rod were cast, one by an ordinary candle, and the other by an incandescent lamp placed some distance further away. The man could work to any candle-power he desired by altering the distance from the candle to the screen. It was his duty to maintain the two shadows of equal intensity.

Gordon soon turned his attention to central station work, and in 1883 he read a paper before the Society of Arts, describing in considerable detail his plans for a station to serve 10,000 lamps, which he was hoping to establish in London. His opportunity to engage in the new industry was not long in coming. In the same year the Great Western Railway Co. came to an arrangement with the Telegraph Construction and Maintenance Co., by which the latter should construct a power station to serve the Paddington terminus and the neighbouring property of the railway, and Gordon was entrusted with the work. In this he was assisted by Mr Frank Bailey who had been associated with him in his electrical work at Greenwich, and the building of the large alternators which had been decided on was undertaken by the Greenwich factory. The Telegraph Construction and Maintenance Co. were to operate the station in the first place, and to supply electricity to the Railway Company under contract, delivery being due to commence on 21 April 1886.

The purpose of the station was to provide electric light for the Great Western Hotel, the offices and platforms of the Paddington terminus, the neighbouring goods yards and lines, and the railway stations at Royal Oak and Westbourne Park, all the lighting of which had previously been carried out by coal gas manu-

factured by the Railway Company. The area to be served was
$1\frac{1}{2}$ miles long and 70 acres in extent. The scheme involved the
supply of current for over 4,100 incandescent lamps of 25 c.p.
each, and 100 arc lamps, mostly of 3,500 c.p. The undertaking
was referred to in 1886 by *The Electrician* as being "by far the
largest installation of mixed lighting hitherto made".

The power house was situated between the Paddington terminus and Westbourne Park station, a quarter of a mile from the
former on the south side of the tracks. The main equipment of the
engine-room consisted of three Gordon alternators driven
directly by vertical engines built by Messrs J. and G. Rennie, one
of the old firms of marine engineers on the banks of the Thames.
Each engine had two cranks, each driven by a 14 in. H.P. cylinder
and a 23 in. L.P. cylinder in tandem, with a stroke of 18 in. It
developed 600 I.H.P. at 146 R.P.M.; working non-condensing with
steam at 160 lb. pressure. The alternators to which these engines
were coupled were of sufficiently imposing dimensions to be
described as "Brobdignagian machines" in the technical press of
the period. They weighed 45 tons apiece, and stood on bed-
plates 18 ft. long. Unlike the design developed almost simultaneously by Ferranti, they had stationary armatures and rotating
fields. The latter, which were 9 ft. 8 in. in diameter and weighed
22 tons, produced an axial flux through the armature coils by
means of 28 horizontal electro-magnets arranged in a circle on
each side. These magnets had cylindrical cores, 6 in. diameter by
3 ft. long, carrying coils which were held in position by the pole-
pieces. The central disk supporting the field magnets was built
up of butt-riveted segments of boiler plate, and was stiffened
by cones of the same material on each side. The armature coils
were wound on bobbins surrounding cores of boiler plate bent
into an acute V shape so as to conform to the segmental shape of
the coils. The cores were bolted to heavy flanged rings of cast
iron supported from the bed-plate on each side of the machine.
They projected inwardly to face the magnet poles carried by the
central wheel. The free ends of the armature cores were fitted with
flanges of german silver perforated with numerous slots so as to
minimize eddy currents. There were 56 armature coils on each

PLATE III. PADDINGTON POWER STATION, 1886

From contemporary engraving in *The Electrician*

side, or twice as many as the number of magnetic poles, so that the machines were really two-phase alternators. The connections to the two phases were, however, kept entirely distinct, each phase supplying current to a separate circuit independently of the other phase. This was not an entirely new idea, for Gramme had previously constructed alternators with as many as eight phases, each to feed an independent circuit of arc lamps. Gordon, however, took advantage of the system to provide greater security, for by connecting his lamps alternately to each circuit he ensured that, in the event of the failure of either circuit, every other lamp, whether arc or incandescent, would be left alight.

In the Gordon machines, two successive coils of each phase were connected in series, all the pairs of coils thus formed being connected in parallel to the machine switchboard. On this board were a number of two-way switches, fourteen per phase, which were used to regulate the output of current in accordance with the load on that phase. Each switch served to short-circuit its pair of coils, and thus to prevent them from supplying current. As may well be imagined, this method of regulation was not conducive to cool running, and to carry away the heat generated a system of water cooling had to be provided. Each of the alternators had a capacity of 350 k.w. Current was generated at 150 v. and about 68 cycles per second. The alternators were excited at 110–130 v. by three 25 k.w. Crompton dynamos driven directly at 400–500 r.p.m. by three-crank tandem compound Willans engines of the earliest design. At full load, each alternator required 85 amp. of exciting current. On a table near the exciters were three ammeters, one for each exciting circuit. The speeds of the exciters were regulated by their stop-valves, which could be operated by a man at the main engines. He knew, by experiment, the exciting current required at various loads, and was guided in his actions by the ammeters mentioned as well as by a pair of voltmeters placed above the machines. On bright days, when not more than 150 lamps were being used, the large alternators were not run at all, current being supplied to the mains by two of the exciters connected in series. In such cases the output was measured by a shunted ammeter. The output of each alternator was measured by an electro-dynamometer in circuit with one of the armature coils, the reading being multiplied by the number of the coils.

The alternators were protected by fuses in the connections to the machine switchboards. From these boards cables were led to the main switchboard situated on a gallery over the exciters. This board was 18 ft. long by 7 ft. 6 in. high, and consisted of five panels of slate. Each half of the board was apportioned to the control of one of the phases. Twelve mains left the station, one pair going to each of the six distributing centres of the system, which were connected by telephone to the power plant. The mains were laid underground, and each pair constituted two separate circuits, each carrying half the lamps for security in case of breakdown. Both circuits could be fed by the same alternator, one of its phases supplying each, or they could be fed by different machines. Again the exciters could be connected to the mains as mentioned above, when the load was insufficient to justify running one of the alternators. To change over from one alternator to another, the incoming machine was run up to speed, and its exciter adjusted to give the same current as that exciting the fields of the outgoing machine. Two men at the switchboard, acting in unison, then opened one switch and closed the other, the machines never being allowed to run in parallel. When the men became expert, this manœuvre could be carried out with hardly a noticeable blink of the lights. The switch blades were of laminated copper strip, built up into bars about an inch square, which were well jammed home into their contacts. When the load on either phase became too great for one machine, feeders had to be switched off and connected to a second machine, for there was no other way of sharing the load.

The engine-room was provided with a double roof and double doors, while each alternator was totally enclosed in a sort of house with double walls of wood and felt to deaden the noise. These houses were fitted with double glass windows to let light into the interior. The machines inside were ventilated by the fan action of their revolving field magnets, the air being discharged through

PLATE IV. BOILER HOUSE OF PADDINGTON POWER STATION, 1886

From contemporary engraving in *The Electrician*

gauze screens on to the railway alongside the power station. The enclosure of the alternators and other precautions against noise were afterthoughts, prompted, no doubt, by the desire to pacify the residents in Gloucester Crescent, some of whom had proceeded, in the winter of 1885, as an infuriated deputation to lay their complaints before Mr de Rutzen, the magistrate at the Marylebone Police Court. Their grievance, as they explained it, was that the Great Western Railway Co. had recently erected machinery of so powerful a character for the lighting of their terminus and adjacent premises that houses in the neighbourhood were almost uninhabitable. The plant, they said, was running night and day, including Sundays, and "the tremendous vibration and noise, added to the fumes of smoke and steam, and the dirt caused by the machinery, produced such a nuisance as to be almost unbearable". The indictment of the power station was, it must be admitted, a fairly comprehensive one, and coming from people accustomed to living in the immediate vicinity of a railway terminus, it is probable that there were some grounds for complaint. The deputation failed, however, in their application for a summons against the Railway Company, the magistrate telling them that before he could grant this, they must bring evidence to show that a danger was being caused to health. Later on they succeeded in obtaining an injunction against the Railway Company, and followed this up in 1887 by an application to the Courts for a writ of sequestration to be issued against the Company for alleged breach of the injunction. This was refused, the Judge saying that he had visited the locality and had found that there was no appreciable discomfort or nuisance caused by the power station. All the satisfaction obtained by the plaintiffs was to be allowed their costs.

The boiler-house contained nine large locomotive boilers supplying steam to a ring main. Not more than five or six of the boilers were required to be in service at the same time, so that there was ample opportunity for the boiler cleaning that was necessary owing to the hardness of the feed water. The supply of steam, moreover, had to be continuous, as the station was never shut down except for a period of three hours every Sunday to enable testing work to be carried out. The boilers were normally fed by a pump delivering water previously heated by exhaust steam in an open-type heater. Alternatively, they could be fed with cold water by injectors, or by an hydraulic accumulator loaded to 180 lb. per sq. inch. Water was supplied to the accumulator by a pump, the speed of which was automatically controlled by a throttle valve linked to the ram of the accumulator, so that the valve was opened and closed in accordance with the movements of the ram. The greater part of the exhaust steam was at first collected by a ring main, and turned, together with the flue gases, into the two 90 ft. chimneys that served the station. Later on, a jet condenser was fixed outside the engine-room immediately behind each engine. This condenser was not provided to create a vacuum for its engine, but served merely to reduce the great volume of steam otherwise discharged into the chimneys, and so to obviate trouble on that score.

Simple as were some of the features of the Paddington installation, others now appear to us as extraordinarily complicated. The lamps, for example, were run at three different voltages, namely 145 v. at the generating station and at the Royal Oak railway station, 120 v. at the passenger station, hotel and goods station, and 105 v. at Westbourne Park railway station. Pilot wires were brought back from the various distributing centres to lamps at the switchboard, which indicated the voltage to the operator by their brightness. The cables serving the distribution centres were known as "divided mains", and their use was one of the most remarkable characteristics of the system. Each of the divided mains was composed of a number of small insulated wires, varying from 21 to 42, according to the feeder. These were connected together into a single conductor at the distributing centre. The wires at the power station end of the main were connected to separate contacts on a "divided main regulator" of which there was one for each main. These were situated in a room called the "wheel room" behind the main switchboard. Each regulator contained a large wheel carrying metal segments round its circumference, which would cover more or less of the contacts according to its angular position. The function of the regulator was to vary the effective number of wires in the main, and thereby to alter its resistance in such a way as to maintain the desired voltage at the distribution centre. The apparatus was operated hydraulically and controlled by solenoids, so that it acted automatically, though it could be worked mechanically by the switchboard attendant. The switching out of a number of lights, for example, would tend to cause a rise in voltage, which the regulator would counteract by reducing the number of wires carrying current in parallel. On the other hand, when more lamps were brought into use, the tendency of the voltage at the distribution centre to fall would be counteracted by an increase in the effective area of the main. The action of the regulators enabled the voltage to be kept within plus or minus 2% of its nominal value, but their response to altered conditions was not very quick, and why such a method was adopted, when the use of an ordinary rheostat connected in series with a main of the ordinary type would have produced exactly the same result, must remain a matter for wonder.

The alternators gave very great trouble before they could be made to run properly. Their mechanical design was excellent, no doubt largely on account of the assistance received from the Marine Engineering Department of the Telegraph Construction and Maintenance Co., but, as regards their electrical features, it must be remembered that Gordon and his assistants were endeavouring to make a great advance into an unexplored field. One of the faults committed—and it was a perfectly pardonable one at the time—was the winding of the armature coils on solid cores of boiler plate. When the first alternator was tested at Paddington, all the useful electrical output that could be obtained was only sufficient to light two or three pilot lamps, although the Rennie engine was developing over 700 I.H.P. at the time! The heating of the armature was so great that one or more coils were burnt out on every trial run. To get over this difficulty all the coils were returned to Greenwich. Longitudinal slots were cut in each core for the accommodation of water pipes, and the machine was provided with a complete system of water circulation to carry away the heat. This was found to make a considerable improvement, but laminated cores were introduced shortly afterwards. Their adoption made the Gordon alternator a success, and burnt out coils became a thing of the past.

The divided main regulators worked, on the whole, automatically and well, but the divided mains themselves were the source of much trouble and expense. As originally made by the Telegraph Construction and Maintenance Co., their individual wires were insulated with gutta-percha, according to submarine cable practice. The large mains leaving the generating station for Paddington were of too great a cross-section for it to be practicable to carry them in the thin buried iron troughs which could be used elsewhere. The vibration caused by the traffic was too severe for iron troughs of a size that could conveniently be handled, especially as, owing to want of space, it was necessary to lay the troughs close to the ends of the sleepers. Wooden troughs therefore had to be used. These stood the vibration, but they did not stand the water blown from locomotive cylinder petcocks, which often soaked the filling of the cable trenches. The action of the "divided main regulators" also was liable to cause overheating of individual wires and melting of their gutta-percha insulation. This combination of circumstances sometimes led to the firing of the insulation, and the burning out of the whole cable, as the fire always started at the damp bottom of the trough. The quality of the material is shown by the fact that when, in 1905, the mains were replaced by others with v.i.r. insulation, the gutta-percha stripped from the wires still commanded a high price. The last of the original divided mains was not done away with before 1906–7, when these mains were superseded by a distribution system from substations fed from the new Great Western Railway Power Station at Park Royal, some of them having worked with gutta-percha insulation for 20 years.

From the lines quoted at the beginning of this chapter, it might be inferred that the arc lighting of Paddington was hardly of a quality to commend itself to the sensitive soul of a poet. The lamps employed were of the Crompton-Burgin type, and as the carbons obtainable were none too good, the lamps required constant attention when burning. To reduce the voltage to that required

across the arc, resistance coils were used, and these gave trouble
owing to the frequency with which they burned out. The incandescent lamps, also, would not have been considered very satisfactory if judged by modern standards. It was the practice to test
every one of them by a photometer before use, and it is recorded
that about 80% of them failed to come within the limits of
tolerance specified. All but the very worst, however, had to go
into service, and their lives were variable and brief. But, even if
there may have been occasional complaints about the light or the
voltage, there can be no doubt that the illumination by electricity
was a great success, and a vast improvement over the gas lighting
which it superseded.

The Paddington Power Station commenced regular service on
21 April 1886, and was operated by the Telegraph Construction
and Maintenance Co. with Mr Frank Bailey as resident engineer,
until towards the end of 1887 when it was taken over by the Great
Western Railway Co. and put in charge of Mr C. E. Spagnoletti,
their chief electrical engineer. It continued in commission until
1907, when it was superseded by a new station erected by the
Railway Company at Park Royal. It thus had a useful life of
about 20 years, which reflects great credit upon Gordon and the
other pioneers with whom he was associated in its design and
construction.

CHAPTER IV

HIGH-TENSION DIRECT-CURRENT SYSTEMS

Confess the failings as we must,
The Lion's mark is always there.

F. T. PALGRAVE

IN the early days of the power station industry, the employment of batteries of accumulators by means of which the constancy of supply could be maintained in case of a temporary breakdown of the generating machinery was much advocated. The system had the advantage of permitting the machines to be shut down altogether during the hours of light load, the supply being then carried on by the batteries. Furthermore, by placing the latter in substations located near the centres of greatest demand, a saving could be effected in the cost of cables, as the mains between the power station and the substations need then be of no greater capacity than that required to carry a small and uniform charging current.

THE COLCHESTER UNDERTAKING

A logical development of the simple substation battery system was to arrange that the batteries should be charged in series and discharged in parallel, thus further reducing the cost of the mains by reason of the higher voltage of the charging current. Although the practice involved somewhat formidable complications, its apparent advantages appealed to many engineers, and a system of electricity supply based on the principle in question was established in Colchester by the South Eastern Brush Electric Light Co., Ltd., in 1884, current being available to consumers on June 11 of that year. The generating station was situated in Culver Street, and the shopkeepers of the High Street and Head Street were expected to furnish a paying load at the commencement. The power plant consisted of a semi-portable Davey Paxman engine developing 90 I.H.P. at 133 R.P.M. The 7 ft. flywheel of this engine was belted to a pulley 4 ft. 6 in. diameter on a countershaft carrying pulleys of 5 ft. diameter from which two Brush arc-lighting machines were driven. The latter were of the 40-light size, the largest then made. Each had an output of 9·5–10 amp. at 1,800 v. when running at 750 R.P.M., so that, when both were operated in parallel, a high-tension current of nearly 20 amp. was available for battery charging. To enable them to work satisfactorily in parallel they were separately excited by small 150 v. dynamos. The exciting current normally required was about 10 amp., this being regulated automatically to suit the requirements of the main generators. The regulator consisted of a small reversible electric motor operating a sliding rheostat in the field circuit, and thus putting more or less resistance in series with the field windings. The motor was started in the required direction by the action of a relay.

Brush arc lighters, with their open-coil windings, were not exactly the most convenient machines for the duty to which these were destined, for they produced a pulsating current which was completely interrupted three times in every revolution of the armature. They had the further drawback that they could not be run on open circuit when separately excited, on account of the sparking at the commutator. To get over this difficulty, resistance coils were provided to give an initial load. The method of operation was as follows. On the switchboard were three ammeters, the first indicating the exciting current, the second indicating the current passing through the resistance coils, and the third showing the current in the charging mains. When the machine was started, the switch was in such a position that the whole of the current generated passed through the resistance. As soon as this current reached a value which showed that the machine was generating at full voltage, the operator changed the position of the switch, thereby cutting out the resistance and connecting the machine directly to the charging mains. An incidental advantage of starting in this way was that the engine was loaded gradually, and there was less risk of the belts coming off than if the load had been suddenly applied.

There were six installations of storage batteries on the system, one at the generating station and the other five in cellars underneath shops in the town. The batteries were arranged on cast-iron racks standing on porcelain insulators. An installation was composed of two batteries, one of which could be charged while the other was in service. Each battery consisted of eight groups of cells, giving 60 v. across its terminals. The eight groups were connected in parallel when supplying current to the 60 v. circuits of the consumers, and in series while being charged. The change-over from charge to service connections was effected by a rocking switch, operated automatically by one of the cells. These switches were a special feature of the system. The sixteen terminals of each battery were connected each to two iron cups containing mercury. The cups were arranged, positive and negative alternately, in two rows at the opposite sides of an insulating base. Between the two rows was a rocking bar carrying two sets of prongs, each set corresponding to one of the sets of mercury cups. When the bar was rocked over to one side, the prongs on that side descended into the cups and so connected the eight groups of cells in series for charging. When it was rocked over on the other side, it connected them in parallel for service. The action of the rocking bar was determined by a solenoid. This was excited by current from the battery, switched on and off by contacts according to the state of charge. The master cell, which controlled this operation, had the upper part of its positive electrode covered by an airtight hood extending down below the surface of the acid. The interior of the hood was connected by a small tube to one side of a flexible diaphragm, the motion of which operated the contacts controlling the action of the solenoid. As the battery became fully charged, the increasing pressure of gas in the hood caused the diaphragm to bulge, thus bringing about the change-over of the rocking bar from the charging to the supply position. The momentary interruption of the circuit during the change-over prevented the high charging voltage of the mains ever entering the houses of the consumers.

The charging and supply mains were lead-covered cables laid in brick trenches directly beneath the pavements, with drainage sumps every 50 ft. The trenches measured 7·5 by 9 in. internally. Where roads had to be crossed, the trenches were covered with cast-iron lids in short lengths. Service connections were placed in wooden troughs under the pavements, with a fuse-box at the entrance to every house. Consumers were charged for current at the rate of 0·5d. per lamp-hour, or alternatively 2d. per lamp per 24-hour day in the summer, and 3d. in the winter. There was no public lighting.

The Colchester undertaking was established in the expectation that 2000 lamps would be served, but it is doubtful whether a quarter of this number were ever connected. The disappointment in this respect, coupled with considerable trouble with the batteries, the original wooden cells of which had to be replaced with earthenware cells on account of leakage, soon brought the Company into financial difficulties. By February 1885 the system was pronounced by critics to be an utter failure, and in April of the same year attempts were made to form another company to save the business from disaster. These came to nothing, partly perhaps because the consumers had been promised by the Brush Co. a refund of 75% of the cost of their installations in the event of the discontinuance of the supply. In August the Town Council agreed to the transfer of the Board of Trade licence, under which the Brush Co. had been operating, to the Colchester Electric Light and Power Co., who announced their intention of making the electric light a success in the town. The resuscitation of the undertaking, however, proved beyond their capabilities, and in September 1886 the whole plant, engine, dynamos, cables and batteries, was sold piecemeal under the hammer. It fetched no more than a total of £345, although many thousands had been spent on it, and gas once more reigned alone in Colchester.

There was no further public supply of electricity in the town for more than twelve years. The Corporation obtained a Provisional Order for the establishment of a municipal undertaking in 1893, but they made no haste to provide it, and it was not until December 1898 that they commenced to give a supply. Their station sent out direct current on the three-wire system, the consumers' pressure being 105 v. Its initial equipment consisted of a pair of Davey-Paxman "Economic" boilers and three Peache-Siemens generating units, with an aggregate capacity of 180 K.W.

THE CADOGAN ELECTRIC LIGHTING CO.

The misfortunes of the original Colchester undertaking did not deter others from attempting to give a low-tension supply of electricity to consumers by means of batteries charged in series from high-tension mains. A renewed effort to realize the advantages of such a system was made by the Cadogan Electric Lighting Co. which was registered in March 1887 with a capital of £30,000 to provide a supply of electricity to the residents of Chelsea, South Kensington, Brompton and Knightsbridge in London. Permission was obtained from the Chelsea Vestry for the use of overhead transmission lines, the Company in its turn agreeing to be bound by the terms of the Statutory Order held by the Chelsea Electricity Supply Co., who were also establishing an undertaking in the same area. The latter Company, who had been refused permission to employ overhead lines, considered themselves badly treated, and endeavoured to prevent their rivals enjoying this facility but without success, as the Vestry held that competition between the two Companies on the terms indicated would be to the advantage of the consumers.

The Cadogan Electric Lighting Co. commenced operations in July 1888 with a system of supply that depended upon the installation of a storage battery in the house of every consumer. These batteries were kept charged by high-voltage transmission lines, on which they were connected in series. The charging current was sent out from a power station situated in Manor Street, on the site of which is now Chesil Court near the Albert Bridge over the Thames. It was transmitted by a single-series circuit consisting of a 19/14 overhead cable suspended from a steel cable carried on poles about 70 yards apart. The circuit was seven miles in length and extended as far as Belgrave Square.

Each consumer was provided with a storage battery comprising from 8 to 64 cells, according to the number of lamps to be served. The cells had a uniform capacity of 700 amp.-hours each. Every battery was divided into four groups of cells, three of which were always available for the lamps, while the fourth group was connected to the charging main. The various groups were changed over automatically from the charging to the discharging arrangement, and vice versa, by a motor-operated switch which disconnected one group from the mains and connected up another at intervals of 1, 2 or 3 min. as required. By this device it was assured that each group of cells was charged during the same period of time, and regularly at short intervals. At every change of connections there was a moment when all four groups were connected to the lamp circuit, and to avoid damage to the lamps at such times, a resistance was automatically put in series with them. Ediswan lamps were used, of 16 c.p. each. These, of course, had to be suitable for the particular voltage of each house, so that the lamps on the system ranged from 12 to 96 v., which was an obvious drawback to the principle.

The power house was a building 90 ft. square, with separate offices facing on the street. It was specially constructed as an electric light station, and was designed to contain machinery for the service of 50,000 lamps. Storage capacity was provided for 40 tons of coal, and a 10,000 gallon tank contained a reserve of water. The engine-room equipment in 1888 consisted of three horizontal compound non-condensing engines of the Armington and Sims type, with cylinders 9·5 and 12 in. diameter by 12 in. stroke. Each engine would develop 50 I.H.P. at 300 R.P.M. It was belted to a "Leeds" dynamo with an output of 70 amp. at 500 v. when driven at 800 R.P.M. The dynamos were of the simple horseshoe design with a horizontal yoke overhead surrounded by a single field coil. They were separately excited by current from a battery to avoid the possibility of a reversal of polarity. Two or more could thus be run safely in series to give the voltage required on the charging mains. The generating sets were built by Messrs Greenwood and Batley of Leeds. They were supplied with steam by a pair of Babcock and Wilcox boilers. The feed water was heated by exhaust steam in a contact heater and delivered to the boilers by a pair of Worthington feed pumps. The boiler-house was served by two iron chimneys, 4 ft. 6 in. diameter by 70 ft. high, which also carried away the exhaust steam.

The Cadogan Company did not prosper. By 1890 it had only 25 houses connected to its system, and these were soon reduced to

16 by a rearrangement of its territory. In February 1891 it went into voluntary liquidation. In anticipation of such an event, the New Cadogan and Belgravia Electric Supply Co., Ltd., had been registered on 30 June 1890, with a capital of £1,000. On 6 July 1892 this Company changed its name to the St Luke's Chelsea Electric Lighting Co., Ltd., increasing its capital at the same time to £30,000. In the following December the St Luke's Company made an agreement to purchase the assets of the Cadogan Company for £4,250 in cash and £4,550 in fully paid shares, but the deal was never completed, and on 5 April 1893 arrangements were made with the Chelsea Electricity Supply Co., Ltd., who possessed a successful undertaking in the same district, to take over the whole interests of the Cadogan Company, for the sum of £10,250, of which £7,000 was to be paid in shares. From that time, therefore, the Cadogan Electric Lighting Co., which then had only about 30 consumers on its system, became merged in the Chelsea Electricity Supply Co., Ltd.

THE CHELSEA ELECTRICITY SUPPLY CO.

This Company had been registered as early as November 1884 but did not commence operations for several years after. It earned for itself the distinction of being the first to break the monotonous record of sterility that had characterized Mr Chamberlain's Electric Lighting Act of 1882 by obtaining Statutory Authority for its undertaking. The Chelsea Electricity Supply Co. was responsible for the solitary Private Bill presented to Parliament in 1886, seeking confirmation of a Provisional Order granted by the Board of Trade for the supply of electricity. The Bill was passed as a separate measure. It empowered the Company to lay mains in Chelsea for the supply of electricity within a given area, stipulating, amongst other things, that a supply should be given to all intending consumers within a distance of 25 yards from a main, and that the price of electricity should not exceed £3. 10s. 0d. per quarter for consumers taking up to 84 K.W.H., with a maximum of 10d. per K.W.H. beyond this consumption. There was also to be a penalty of £2 per day per consumer for interruption of supply, and an equal penalty for each public lamp unlit, with the provision that the aggregate penalties should not exceed £50 per day.

Although, as already mentioned, the Chelsea Company obtained statutory powers in 1886, it was unable to secure sufficient financial support to proceed with its scheme at the time. The passing of the amended Electric Lighting Act of 1888, however, with its more liberal provisions as regards compulsory purchase of Company undertakings, made financing possible. Certain manufacturing firms, including Callender's Cable Co., the Electrical Power Storage Co., and the Electric Construction Corporation agreed to take an interest in the fortunes of the Chelsea Company and provided between them a large part of the plant necessary for it to carry out its programme. It was thereby enabled to commence service in 1889.

The system of supply adopted was that known as the Beeman-Taylor-Kine system, the patents of which had been acquired by the Electrical Power Storage Co., Ltd. Its characteristic feature was the supply of low-tension direct current to a distribution network from automatic battery substations, the batteries being charged in series and discharged in parallel. This, it will be remembered, was the principle adopted at Colchester in 1884. The Colchester undertaking had, as already recorded, been a disastrous failure, but its technical defects were considered to be capable of remedy, and it was believed that with more appropriate generating machinery and a more reliable type of battery, the system could be developed along successful lines. It was also intended to improve on the practice at Colchester by the use of "continuous current transformers" or "dynamotors" as they were sometimes called, which would enable the high-voltage current from the generating machinery to be converted to low-tension current which could be delivered to the network in parallel with the current from the batteries at times of heavy demand. The consumers were supplied at 100 v. on the two-wire system. Vulcanized bitumen cables were employed, drawn into Callender-Webber bitumen casing. This kind of casing was frequently employed in the early days of electrical distribution, and it gave good service. It was found, however, to be dangerous under certain circumstances, as it produced an inflammable gas in the event of an arc occurring, and several explosions were due to this cause. It was also liable to be distorted by heat. For these reasons, from 1897 onwards, it was gradually superseded by Doulton earthenware casing, and in 1902 the replacement of the bitumen feeders and distributors by lead-covered cable was begun.

A generating station was established by the Company in the basement of a house in Draycott Place, now known as Cadogan Gardens. The basement was 9 ft. below the street level, and was equipped with four Babcock and Wilcox boilers with a rated capacity of 2,400 lb. of steam per hour, and three generating sets. The latter consisted of Brush-Victoria continuous-current dynamos directly coupled to vertical Brush engines, which were, however, shortly afterwards replaced by Willans engines. Each dynamo had a capacity of 75 amp. at 500 v., the machines being operated in series to give the voltage required for charging the batteries. A separate engine-driven dynamo with compound winding was provided for their excitation, and this machine served also for the station lighting. The engines exhausted into the chimney, but some of the exhaust steam was by-passed through a surface-heater, supplied with water from a tank on the roof.

There were three battery substations, one at the power house and the other two at Pavilion Road and Egerton Mews respectively. Every substation contained a battery consisting of eight groups of 54 E.P.S. cells in glass boxes. The cells stood on sawdust in varnished wooden trays which rested on oil-filled insulators carried on wooden beams. The ends of the beams were held in sockets of cast-iron standards which supported the whole. From eight to ten tiers of cells were carried in this way, according to the height of the substation. The various batteries were charged half at a time, four groups of cells being connected in series with each other for this purpose, and also in series with the corresponding half-batteries in the other substations. The other four groups of cells in each battery were connected in parallel across the low-tension supply mains. It will be remembered that at Colchester there were two complete batteries per substation,

PLATE V. BOILER HOUSE OF DRAYCOTT PLACE POWER STATION, 1889

of which only one was discharging at a time. At Chelsea, on the other hand, each substation contained only one battery, the two halves of which could both be discharging into the supply mains in parallel at times of heavy load.

The method of operating the plant was as follows. At about 4·0 a.m. every morning, after getting up steam, the engineer in charge of the power station closed a switch which put the first half of every battery in series on the charging mains. The aggregate voltage of the batteries was then indicated by a voltmeter. The generators were then run up, connected in series and, after adjustment to give the required voltage, were switched on to the charging mains. By about 10.0 a.m. the half-batteries would be charged. Each of them was provided with a "master cell" as at Colchester, to control the automatic change-over mechanism, but this apparatus was much more elaborate than in the earlier installation. The hydrogen gas given off at the end-plate of the master cell was collected in a hood of ebonite covering the top of the plate, as in the older arrangement. It was led thence by a tube to a sort of gasometer which was provided with a relief valve operated by the charging current. When the master cell of the first half of a battery was gassing freely, indicating that the remainder of the cells were fully charged, the gasometer rose, and in doing so operated switches which changed the first half of the battery from the charging mains to the supply mains. To maintain the continuity of the charging current, it was shunted through a resistance during this operation. The cessation of the charging current in the battery circuit caused the relief valve on the gasometer to open, letting the gas escape. The gasometer therefore fell, and in its descent it operated switches that disconnected the second half of the battery from the supply mains and connected it to the charging mains. The resistance across these mains was then cut out, and the battery was left with the second half being charged and the first half supplying the demands of consumers. By about 4.0 p.m. the charging of the second half would be complete, and the rise of the gasholder would remove it from the charging mains and place it in parallel with the other half across the supply mains.

The regulation of voltage on the supply mains was effected automatically by the employment of four counter-E.M.F. cells. A solenoid was connected across the mains and provided with a core which rose and fell in accordance with the voltage. The core, by its motion, operated contacts, and thereby brought about the addition or the removal of counter-E.M.F. cells, until the desired voltage was restored. By this means the voltage of supply was maintained within a range of plus or minus about 1·5 v.

The automatic arrangements on the Chelsea system were much more complicated than those adopted at Colchester, but otherwise the equipment was superior in most respects. The generators were more suitable for battery charging than the Brush arc-lighting machines, and the batteries, too, were immensely superior to the Planté batteries with their thin sheet-lead plates and their generally delicate construction, which had been used at Colchester.

The first continuous-current transformer was installed about the end of 1889, and proved so successful that within a year or two all the battery stations were provided with them. They were built by the Electric Construction Co. of Wolverhampton, and had the general appearance of an overtype bipolar dynamo with a commutator at each end of the armature. The latter carried a double winding. The high-tension current from the main generators was passed through one winding, and drove the machine as a motor, while low-tension current was generated in the other winding. Each transformer had a capacity of 40 K.W. at a speed of 550 R.P.M. and its efficiency was said to be 92% at full load. The weight of the machine was about 5 tons, and the floor space occupied measured 7 by 3 ft. The object of the transformers was to enable the 500 v. generators to supply 100 v. current to the mains, and thus to assist the batteries during the peak of the load. To start up a transformer in a substation, one of the dynamos at the power station was connected to the high-tension brushes of the transformer through the ordinary battery charging mains. The dynamo was then run up, with a resistance in its field, until it showed the required voltage at its terminals. This caused the transformer to develop a voltage on its low-tension side equal to

PLATE VI. FLOOD STREET POWER STATION, 1921

By courtesy of *Central London Electricity Limited*

that in the circuit into which the batteries were discharging. The transformer could then be switched in on the low-tension side. The current in the high-tension circuit was next increased by the gradual removal of the resistance in the fields of the charging dynamo, until the maximum capacity of the transformer was attained, after which the transformer would continue to work at this capacity, all variations of load on the system being taken care of by variations in the discharge current of the batteries.

After the purchase of the undertaking of the Cadogan Company in 1893, which has already been mentioned, the Chelsea Company felt itself free to proceed with plans for the systematic development of the larger area now practically under its control. It is true that the London Electric Supply Corporation had powers to supply alternating current in the same territory, but this Company apparently had its hands too full with the development of its business in other quarters, not to mention its troubles with the Deptford Station, to be much concerned with waging war with the Chelsea Company on its own ground. At any rate its competition was never severe, and after the disappearance of the Cadogan Company the Chelsea Company enjoyed a virtual monopoly in the area. In conformity with a comprehensive scheme for the service of the district, it immediately reorganized the Manor Street Power Station. The Armington and Sims engines with their belt-driven dynamos were removed and replaced by Willans-E.C.C. direct-coupled units, and the station arranged to run in parallel with the main station of the Company at Draycott Place. At the same time compulsory powers were obtained for the purchase of the site of a new generating station in Flood Street, and for new substations at Elm Park Gardens, Claybon Mews and Pond Place.

The Flood Street Station, built in 1894, commenced operations with two 80 k.w. Willans-E.C.C. direct-current generating sets supplied with steam from a pair of Babcock and Wilcox boilers. Its capacity was soon increased by the installation of larger units of the same type, and in 1898 it took over the load of the plant at Draycott Place which then ceased to serve as a generating station. By 1911 the steam-driven plant at Flood Street comprised 14 units aggregating 2,800 K.W., this total being made up of two units of 80 K.W., four of 150 K.W., six of 200 K.W. and two of 420 K.W. capacity each. All were of Willans-E.C.C. manufacture. Some generated at low voltage for the local network and the others at 1,500 v. for the charging of the batteries at the three substations in series. In 1911 a new departure was made by the installation of three 200 K.W. dynamos driven by Diesel engines in the Flood Street Station, bringing the number of units up to 17, with a total capacity of 3,400 K.W. This equipment met the needs of the Company until 1922, when arrangements were made to take a bulk supply of electricity at 6,600 v. from the new Grove Road Station of the Central Electric Supply Co. The steam plant at Flood Street was then shut down and replaced by a number of motor-converters, but the Diesel sets were retained till 1928, when the consolidation of the principal London electrical under-takings by the London Power Company enabled the Chelsea Company to obtain a bulk supply at 22,000 v. and to give up generating work altogether.

As regards other features of the Chelsea undertaking, the original two-wire supply to consumers at 100 v. was changed in 1897 to a three-wire supply at 200/100 v. to enable distribution costs to be reduced, and to take advantage of the 200 v. incandescent lamp which had then been recently introduced. Another advance in the same direction was made in 1912 by raising the supply voltage to 400/200, with a further increase in the capacity of the distribution network. The most drastic change, however, had to be made in 1928, when the Company was confronted with the heavy task of converting the whole of its undertaking to the standard alternating current system, a work which has taken ten years to accomplish.

The automatic operation of the batteries, characteristic of the original undertaking, ingenious as it was, could not stand the test of experience. It was found to be practically impossible to leave the substations unattended, and this being so, there was no need for the complicated and unreliable automatic gear. The batteries were retained for many years, being employed in conjunction with the motor-generators, and allowing the latter to be shut down entirely at times of light load. In due course, how-ever, the batteries themselves were discarded, and the substations operated as simple motor-generator stations. The elimination of the batteries obviated the necessity for the 1,500 v. charging current, and the transmission voltage was then reduced to 1,000 v. for the motor-generators.

THE OXFORD SYSTEM

The Oxford Electric Co. was registered in August 1891 with a capital of £100,000 to take over the Electric Lighting Order granted by Parliament the previous year to the Electric Installa-tion and Maintenance Co., Ltd., for the establishment of a public supply of electricity in Oxford. The Local Authorities, in giving their sanction for the undertaking, had stipulated that a direct-current supply should be provided. The best available site for a generating station was, however, about three-quarters of a mile from the centre of the load, and a purely low-tension system would have therefore involved a heavy outlay on mains. Under the circumstances it was decided, on the advice of Mr Thomas Parker the chief engineer of the Electric Construction Corpora-tion who designed and carried out the installation, to adopt what afterwards became known as the "Oxford System". This system, which had already been tried on a small scale at the Crystal Palace, was characterized by the use of direct current generated at a fairly high voltage for transmission to substations where it was converted to low-tension current for distribution. The conversion was effected by rotary motor-generators, or as they were then called "continuous current transformers". The principle was, in fact, analogous to that of an ordinary alternating-current system, with high-tension transmission lines and step-down transformers in substations, but it possessed the advantage that storage batteries could be employed, and the generating machin-ery therefore could be shut down altogether at times of light load. The Oxford system resembled, in some respects, that of the Chelsea Electricity Supply Co., but was much simpler and more practical by reason of the absence of any attempt to control the batteries by automatic devices.

A site was acquired by the Company at Cannon Wharf, Osney, on the banks of the river Isis, where water was available for condensing purposes and operations could be carried on without causing a nuisance. Since it falls to so few electrical undertakings to inspire the pen of a poet, the following verses relating to the Oxford supply may perhaps be quoted. They are taken from an ironic poem by Hilaire Belloc entitled *Lambkin's Newdigate*:

> Descend, O Muse, from thy divine abode
> To Osney, on the Seven Bridges Road,
> For under Osney's solitary shade
> The bulk of the Electric Light is made.
> Here are the Works, from hence the Current flows
> Which, (so the Company's prospectus goes)
> Can furnish to subscribers, hour by hour,
> No less than sixteen thousand candle power,
> All at one thousand volts...
>
> .   .   .   .   .
>
> As for the lights they hang about the town,
> Some praise them highly, others run them down.
> The system (technically called 'The Arc')
> Makes some passages too light, others too dark,
> But in the house the soft and constant rays
> Have always met with universal praise.

In amplification of the information contained in the above lines, it may be said that the works in question consisted of a brick-built power house measuring 80 by 60 ft., which was equipped with three boilers of the locomotive type and three main generating units. Each of the latter consisted of a vertical triple expansion engine, by J. and H. McLaren of Leeds, driving an E.C.C. dynamo by means of a link belt from its flywheel. The engines were of the open-fronted marine type, with cylinders 9, 14·25 and 22·5 in. diameter with a stroke of 2 ft. The H.P. and I.P. cylinders were jacketed with live steam, the jackets draining to the hot-well through steam traps. The engine speed was 125 R.P.M., this being maintained by a flywheel governor which varied the cut-off in the H.P. cylinder. Behind each engine was its surface condenser, with 382 sq. ft. of cooling surface, and behind this again was an air pump 11·5 in. diameter, a circulating water

PLATE VII. OXFORD POWER STATION, 1892

From contemporary engraving in *The Electrical Review*

pump 10 in. diameter, and a feed-pump 1·75 in. diameter, all with a stroke of 14 in. These pumps were actuated, as in marine practice, by levers from the cross-head of the I.P. cylinder. The engines, as will be gathered, were of a highly economical type, and were reputed to be capable of working with a consumption of 1·3 lb. of coal per I.H.P.-hour. The direct-current generators which they drove ran at 400 R.P.M., and had each a rated output of 80 amp. at 1,080 v. They were separately excited by small Elwell-Parker dynamos driven from rope pulleys on the armature shafts of the main sets. The exciters worked at 135 v., so that current could be taken from them for charging the accumulators used for lighting the station. The main units were operated in parallel, and their current left the power house by two pairs of 37/14 V.I.R. high-tension cables drawn through cast iron pipes.

The high-tension current was delivered to a switching station in Broad Street, approximately at the centre of the load area. At this station there was installed a continuous-current transformer taking in current at 1,000 v. and having an output of 360 amp. at 110 v. on the low-tension side. The machine was of the bi-polar type with a double-wound armature having a commutator at each end. In series with the high-tension armature winding were a few turns round the field magnets to give sufficient excitation to start the machine. The main excitation was due to a low-tension shunt winding. The armature ran at 550 R.P.M., and its bearings were continuously supplied with oil by a pump driven from the end of the shaft. Besides the transformer, the switching station contained two storage batteries, each capable of an output of 120 amp. for eight hours.

Substations, each equipped with a transformer similar to that described above, and connected by high-tension mains to the switching station, were established in Queen Street and King Street, and subsequently in other localities. Their machines were started, controlled and stopped by the operator in the switching station. They were started by closing a double-pole switch on the high-tension feeder, in series with which was a rheostat to check the first rush of current. As the resistance was cut out, the low-tension voltage of the transformer rose to that of the distribution system, and the machine could then be switched in on the low-tension side. The low-tension voltage of the transformer was shown by a voltmeter at the switching station. In one of the leads of this voltmeter was the coil of a relay, which was unaffected because of the smallness of the current which could pass through the resistance of the voltmeter. When the correct voltage was obtained, the attendant momentarily short-circuited the voltmeter. The current that could then flow in the circuit was sufficient to operate the relay in the substation, and thus to bring about the closing of the switch. The voltage supplied by the transformer was then regulated by the rheostat in the high-tension mains at the switching station. To make this regulation as effective as possible there were voltmeters in the switching station connected by pilot wires to each of the low-tension feeding points.

The Oxford Station commenced supply on 18 June 1892, the day before Commemoration Week. Five 3,000 c.p. arc lamps had been erected at important places in the City, and several of the Colleges and numerous other consumers had been connected up. At the end of 1893 the connections were equivalent to 7,012 lamps of 8 c.p., and 103,000 K.W.H. had been supplied to consumers during the year. Of this quantity, 18% had been delivered by the batteries, the generating station being shut down meanwhile. The employment of batteries, moreover, brought about another economy, for their availability made it unnecessary ever to run a transformer at less than half-load, at which output its efficiency was still comparatively high. Indeed, it is said that after making allowance for the losses in the batteries and transformers, not less than 62% of the power generated was delivered at the terminals of the consumers.

By 1903 the capacity of the generating plant at Oxford had been increased from the original 240 K.W. to a total of 850 K.W., the machinery then consisting of five triple-expansion McLaren engines driving E.C.C. dynamos by belts, and one 225 K.W. Willans-E.C.C. direct-coupled set. Steam was supplied by three McLaren locomotive boilers, two Davey Paxman locomotive boilers, and two of the latter firm's "Economic" dry-back marine type. A few years later the addition of more direct-coupled units brought the capacity of the 1,000 v. continuous-current plant to about 1,500 K.W., and then alternating current was introduced by the installation of a three-phase Belliss-E.C.C. set of 80 K.W. capacity generating at 3,000 v. and 50 cycles. A motor convertor allowed this machine to supply current to the 100 v. direct-current network when desired. It was also linked to the 1,000 v. continuous-current system by means of a motor generator, so that it could supply the high-tension direct-current mains as well as performing its normal duties.

The so-called "Oxford System" of generating direct current at 1,000 v., and reducing the voltage to the network by means of continuous-current transformers or motor generators, was also adopted by the Crystal Palace District Electric Supply Co. in 1893, by the Charing Cross Electricity Supply Co. in 1896, and by the Municipality of Shoreditch in 1897. Reference to the latter undertakings will be found elsewhere.

THE WOLVERHAMPTON HIGH-VOLTAGE DIRECT-CURRENT SYSTEM

Another example of high-voltage continuous-current working was that of the Wolverhampton Municipality. This followed generally along the lines of the Oxford undertaking, but was noteworthy for the fact that a working pressure of 2,000 v. was adopted, in place of the 1,000 v. employed at Oxford. The Wolverhampton Power Station, erected in the Commercial Road on the banks of the Canal, was formally opened by Lord Kelvin on 30 January 1895, although current had been supplied from it about a fortnight previously. It contained three main generating units, two of them consisting of 140 K.W. dynamos driven at 400 R.P.M. by 15 ropes from the 14 ft. flywheels of Marshall compound side-by-side engines with cylinders 14·5 and 25 in. diameter by 32 in. stroke. The smaller dynamo had a capacity of 65 K.W. at 500 R.P.M. and was driven by eight ropes from the 9 ft. flywheel of a Marshall engine with cylinders 10 and 17·5 in. diameter by 24 in. stroke. The engines took steam at 160 lb. pressure from three Lancashire boilers, and exhausted into a Ledward ejector condenser. All three dynamos were of the E.C.C. manufacture and generated current at 2,000 v. From each of the two larger ones an exciter was driven at 720 R.P.M. by three ropes, and there was, in addition, a spare exciter driven by a Bumsted and Chandler engine. All the dynamo bearings were supplied with automatic forced lubrication by small plunger pumps driven by eccentrics turned in the shafts.

The dynamos, which were operated in parallel when required, delivered 2,000 v. current to a central switching station in the basement of the Town Hall, from whence the machinery in the three substations was controlled. One of the substations was at the Town Hall itself. This contained two 45 K.W. motor generators to reduce the pressure to 110 v. for the two-wire distribution network, one arc-light balancing motor generator with 1,000 v. across each commutator, and a booster for charging batteries off the low-tension mains. Two other substations—at the Art Gallery and the Free Library respectively—each contained a 45 K.W. motor generator reducing the voltage from 2,000 to 110 v. There were also 110 v. batteries in the substations which served to carry the load after the generating station had been shut down at midnight.

All motor generators were started and stopped by remote control from the central switching station. They fed into a common two-wire network. Each pair of trunk mains had a capacity of 120 amp. at 2,000 v. Vulcanized bitumen cables were used, laid on wooden bridges in cast-iron troughs which were afterwards filled in with bitumen. For the public lighting there were forty 2,000 c.p. arc lamps. These were arranged, twenty in series, on each side of a three-wire system which was kept in balance by the machine already referred to.

The Wolverhampton undertaking was extended from time to time, and the capacity of the 2,000 v. generating plant eventually reached about 1,100 K.W., but the system suffered from a number of breakdowns due mainly to the high voltage, and in 1899 it was decided to change over gradually to the more reliable three-wire direct-current system with 440 v. across the outers.

CHAPTER V

SOME OF THE EARLY LONDON
SUPPLY COMPANIES

What shocks one part will edify the rest,
Nor with one system can they all be blest.   POPE

ALTHOUGH notable examples of early electrical enterprise were given by a few provincial towns such as
Brighton, Eastbourne and Hastings, all of which started a
public supply of electricity in 1882, London was naturally the principal centre of electrical development. Many Supply Companies
were operating in the Metropolis by the end of 1891, at which
date such important centres of population as Manchester, Leeds,
Edinburgh, Hull, Nottingham, Leicester, Wolverhampton and
Norwich had not even begun the construction of electric light
stations. The London Companies, moreover, provided between
them examples of every known system of generation and distribution, so that a brief review of the principal undertakings started
in the Metropolitan area during the eighties and nineties of
the last century will give a fair idea of the state of the art at
the time.

Of the London undertakings, those of the London Electric
Supply Corporation, the Cadogan Electric Lighting Co., and
of the Chelsea Electricity Supply Co. have already been described
in some detail. Others will now be considered in turn.

THE METROPOLITAN ELECTRIC SUPPLY COMPANY

After the completion of their contract with the Great Western
Railway Co. for the construction and operation of the power
station at Paddington, described in Chapter III, the Telegraph
Construction and Maintenance Co. decided to discontinue their
activities as regards power station work. They had spent an
enormous amount of money on the Paddington installation, and
although they had the satisfaction of knowing that they had

carried it to a technical success, they no longer felt disposed to proceed with a scheme for building a station three or four times the size in the St James's district, for the lighting of which they had obtained a concession. Gordon and Bailey, however, retained their enthusiasm for central station development, so they severed their connection with the Telegraph Construction and Maintenance Co. early in 1888, to continue the sort of work they had found so congenial. The opportunity was offered them by the construction of Whitehall Court, which was then being put up by the ill-fated Liberator Building Society. Jabez Balfour, with Gordon, Bailey and others, formed a Company called the Whitehall Electric Supply Co., Ltd., which was registered in October 1887 with a capital of £200,000, the object of the Company being to build a power station to light Whitehall Court and several large buildings in the immediate neighbourhood.

Gordon's experience at Paddington had convinced him that alternating current was definitely not the best form of supply for lighting work, and although he had been one of the great protagonists of that system, he publicly renounced his earlier opinions and became an advocate of a direct-current supply, the continuity of which could be assured by means of batteries. This, therefore, was the system employed for the Whitehall undertaking. The power station was built underground beneath what is now the roadway in front of Whitehall Court, as no other site was available for it. The excavation, which had a useful area of 3,500 sq. ft., was roofed over by steel decking to carry the roadway. The station consisted of a boiler-room, engine-room and battery-room. In the boiler-room were three double-furnaced wet-back marine boilers built by Messrs Hick, Hargreaves and Co., Ltd., of Bolton. They were 9 ft. in diameter by 8 ft. long, working at 140 lb. pressure, and each was capable of evaporating 3,500 lb. of water per hour. The main flue had to make various bends, and at one place made a drop of 12 ft. in order that it might pass underneath the foundations of Whitehall Court on its way to the 135 ft. stack from which the gases were discharged. It is interesting to note that these same three boilers, with the makers' name and the date 1888 on their cast-iron furnace fronts, are still (1939) in regular service, supplying steam for the purposes of Whitehall Court.

In the engine room were three 170 I.H.P. Willans engines running at 350 R.P.M. and directly coupled to 110 V. continuous-current dynamos. The latter were, at first, Crompton machines, but these were replaced by dynamos of Siemens make, and the Crompton-Howell storage battery gave way at the same time to a battery of Elwell Parker manufacture. The exhaust from the engines was passed into two large Berryman water heaters, in which most of it was condensed at atmospheric pressure, the excess being discharged through a vertical cast iron pipe 16 in. in diameter, erected inside the chimney. One of the Berryman heaters was used for heating the feed-water from the mains, which was previously treated in a water softener. The other heater supplied hot water for heating purposes in Whitehall Court.

The battery-room, which measured 43 ft. by 15 ft. 6 in., contained an 800 amp.-hour Elwell-Parker battery with the cells erected in four tiers. There were altogether 960 cells, divided into eight groups, each group consisting of two parallel sets of 60 cells. The voltage of discharge was regulated by sliding switches which varied the number of cells in series. For charging there was an arrangement by which the eight groups of 60 double cells could be transformed into ten groups of 48 double cells.

The station could not be ventilated by ordinary methods, as no opening into the roadway above it was permissible. The draught of the furnaces of course served to renew the air in the boiler-room, but air had also to be changed in the engine- and battery-rooms. From these places the heated air was exhausted by means of a large duct placed just under the ceiling, and furnished with openings at various points. The duct terminated in the flue, so that the chimney draught was used to provide the necessary suction.

The Whitehall Court Station was opened on 5 October 1888, and within a few weeks the Company had customers for all the current the plant could produce. The price charged was 30s. per annum for the equivalent of every 8 c.p. lamp installed, and a service was given to the large hotels in Northumberland Avenue, and up to the Church of St Martin in the Fields, which was also lit by electricity. An existing subway underneath Northumberland Avenue facilitated the laying of the mains, which were buried in wooden troughs in the bottom of the subway. Elsewhere, Callender-Webber bitumen conduits were employed.

After the fire in the Sardinia Street Station of the Metropolitan Company, in 1895, two of the generating sets were removed from Whitehall Court Station to Sardinia Street, and replaced by a pair of smaller sets of about 50 K.W. each, and the battery which was worn out was also removed.

In November 1887 a Company was registered under the name of the South Metropolitan Electric Supply Co., Ltd., with a capital of £250,000, "to produce electric, magnetic or other force, and to supply the same for light and heat". In the following July it decided to drop the restrictive epithet "South" from its title, to increase its capital to £500,000, and it set out on its successful career as the Metropolitan Electric Supply Co. by purchasing the assets and unfinished power station of the Whitehall Company for £40,211, of which £19,891 was paid in cash and the balance in shares. The first chairman of the Metropolitan Electric Supply Co. was Sir John Pender, Mr J. E. H. Gordon was its chief engineer, with Messrs Frank Bailey and C. A. Holbrow, both of whom had been at the Paddington Station from its very start, as his two assistants. Soon after the formation of the Metropolitan Company, Gordon left it to establish himself as a consulting engineer, in which capacity he was responsible for the design of several power stations before he was accidentally killed by a fall from his horse at the age of only forty years. Bailey succeeded him as the chief engineer of the Company.

After acquiring the Whitehall Court undertaking, the Metropolitan Company purchased, in February 1889, also as a going concern, a small power station in Rathbone Place, off Oxford Street, which had been started in September 1887 by Messrs G. E. Pritchett and Co. as a private speculation. The first machine in the station was a 32 K.W. Ferranti alternator generating at 100 V. This pressure was stepped up to 2,000 V. for overhead transmission, and stepped down again on the consumers' premises.

Early in 1888 the Ferranti alternator was replaced by two Mordey alternators, the first of their kind to be made. They were driven by ropes from 80 H.P. engines, and supplied current to about 1,500 lamps by means of transformers. The Metropolitan Company substituted for the Mordey machines a pair of 100 K.W. Elwell-Parker alternators driven directly by two-crank Willans triple-expansion engines, the first of these going into service in November 1889.

The Rathbone Place Station occupied the basement and ground floor of a building 118 ft. long by 44 ft. wide, the intervening floor being removed to give height. When the Metropolitan Company took it over they laid it out for an equipment of five 100 K.W. Willans Elwell-Parker units running at 350 R.P.M. and generating single-phase current at 1,000 v. and 105 cycles. Two separate exciting dynamos of 28 K.W. each, at 140 v., also driven by Willans engines, served the main generators. In the basement of the building were five Babcock and Wilcox boilers, each designed to evaporate 6,000 lb. of water per hour. The feed-water was passed through a Berryman heater, supplied with exhaust steam from the engines, the remainder of the steam escaping to atmosphere. A brick chimney, 120 ft. high, finished externally with white glazed bricks, served the boiler-house.

As a result of the Board of Trade inquiry, in 1889, into the question of the electricity supply of London, the Metropolitan Company were granted Provisional Orders covering large areas in the districts of Marylebone, Bloomsbury, Lincoln's Inn, and Covent Garden. The first power station it established to cope with its responsibilities for the supply of electricity in its area was one in Sardinia Street, near the south-western corner of Lincoln's Inn Fields, the site of the station being now covered by Kingsway. The engine-house and boiler-house were separate buildings of brick, the former being two storeys in height, with the engine-room on the ground floor and the electrical machinery in the room above. On Bailey's advice an alternating-current supply was decided on, and this system was adhered to in all subsequent stations of the Company, although, as will be seen, continuous current crept in later. The machinery for Sardinia Street was obtained from the Westinghouse Co. of Pittsburg, U.S.A., and the station is worth describing in some detail, as it was considered to be representative of the best American practice at the time.

The prime movers consisted of ten Westinghouse compound engines arranged on either side of a central gangway. Five of the engines had cylinders 14 and 24 in. diameter by 14 in. stroke. These were fitted with 90 in. flywheels, and developed 250 I.H.P. at 280 R.P.M. The other five engines had cylinders 16 in. and 27 in. diameter by 16 in. stroke; their flywheels were 110 in. diameter, and the engines developed 300 I.H.P. at 250 R.P.M. Also on the ground floor were three exciter engines with cylinders 10 and 18 in. diameter by 10 in. stroke. These had 60 in. flywheels and developed 65 I.H.P. at 300 R.P.M. From the flywheel of each engine, a belt—20 in. wide for the main units and 14 in. wide for the exciters—was taken upwards through the ceiling at an angle of about 30° to the alternator-room above. On this floor were ten Westinghouse single-phase alternators, each with an output of 125 K.W. at 1,000 V., and a speed of 1,050 R.P.M. They were excited at 100 V. by the three belt-driven dynamos, each of which could supply sufficient field current for six of the alternators. Continuous lubrication of the machinery was carried out by a gravity supply of oil from an overhead tank.

The switchboard was an imposing structure, 54 ft. long by 11 ft. high, standing only 2 ft. from the wall. It comprised 20 panels of enamelled slate mounted in polished wood, and was divided into three sections for the control of the exciters, alternators and feeders respectively. The alternators were never run in parallel, and matters were so arranged that, if the board was operated properly, it should be impossible to put them into parallel inadvertently or otherwise, or to short-circuit one through another. Each alternator had its own pair of bus-bars to which it could be connected by a pair of single-pole switches. It was also provided with a double-pole field switch and a field rheostat. There were twenty feeders, each of which could be plugged on to any one of the alternators at will. Every feeder had its own ammeter, voltmeter and switch, and all feeders were taken from the switchboard, across the station yard, to a fuse-house on the other side,

PLATE VIII. SARDINIA STREET POWER STATION, 1891

From contemporary engraving in *The Engineer*

PLATE IX. BOILERS IN COURSE OF ERECTION AT SARDINIA STREET POWER STATION, 1889

where they terminated in fuse boards from which the distribution cables radiated.

The boiler-house contained twelve Babcock and Wilcox boilers working at a pressure of 150 lb. per sq. in., each having an evaporative capacity of 4,200 lb. per hour. They were placed in an excavation so that the centres of the drums were about at ground level. The reason for this was to facilitate the supply of coal from an underground storage bunker, of 500 tons capacity, at the end of the boiler-house. From this bunker a narrow-gauge railway ran along the boiler-house between the two rows of boilers. Seven of the latter were on one side of the central aisle and five on the other, the remainder of the space being occupied by three Worthington feed pumps and a like number of feed-water heaters. Raw water was used for boiler feed, as the engines were non-condensing.

The cables leaving the power station were drawn into cast-iron pipes from 3 to 5 in. diameter. The current was transformed down to 100 v. for the consumers, who were charged 7·5d. per K.W.H. plus a meter rent of £1 per annum, a main switch rent of 7s. 6d. per annum, and a transformer rent of £1 per annum for the equivalent of every fifty 8 c.p. lamps connected.

The Sardinia Street Station went into commission on 1 September 1889. On 24 June 1895 it was practically destroyed by a disastrous fire which broke out behind the switchboard, and of which the cause was never satisfactorily determined. The conflagration started at 6.15 p.m., and resulted in a complete cessation of supply from the station, but by 8.30 p.m. most of the theatres in the neighbourhood were lit by current from other stations of the Company, though several days had to elapse before a service could be given to all consumers. Both the London Electric Supply Corporation and the City of London Electric Lighting Co., who had mains in the neighbourhood, came to the rescue of their Metropolitan rival, supplying current from Deptford and Bankside respectively.

The fire made an end of the reciprocating engines and their belt-driven alternators at Sardinia Street. Their suppression had already been foreshadowed by an order placed in the previous April for four 350 K.W. Parsons turbo-alternators to run at

3,000 R.P.M. and generate single-phase current at 1,000 v. and 50 cycles, the machines, in fact, being almost precisely similar to the sets which had saved the situation at the Manchester Square Station the year before. Immediately after the fire a fifth unit of the same capacity and speed was ordered, and later on two more, making a total capacity of 2,450 K.W. of turbine machinery, or more than double the output of the reciprocating sets that had previously occupied two floors of the station.

In 1897 the Company decided to convert the station from alternating to continuous current, and in May of that year a start was made by ordering two 250 K.W. turbo-dynamos, each with an output of 1,140 amp. at 220 v. from Messrs C. A. Parsons and Co. These ran at a speed of 3,000 R.P.M. They were followed by a 75 K.W. direct-current set, generating at 110 v. and running at 4,800 R.P.M. Between 1898 and 1902 the conversion of the seven large units to 250 K.W. direct-current generators was carried out, the station then producing direct current only. All the turbines worked non-condensing, as there was no possibility of doing otherwise at Sardinia Street.

After Sardinia Street, the next station of the Metropolitan Company to go into service was that at Manchester Square, which was built to supply the Marylebone area allotted to the Company as a result of the Board of Trade inquiry in 1889. The Manchester Square Station was constructed to the designs of Mr J. E. H. Gordon, and started running in January 1890. It contained ten Willans engines of 200 H.P. each, direct-coupled to single-phase Parker alternators which were driven at 350 R.P.M. There were four exciters, also driven by Willans engines, and steam was supplied at 150 lb. pressure by nine Babcock and Wilcox boilers. The engines worked non-condensing, the exhaust steam being discharged into a cast-iron pipe inside the smoke-stack. Current was generated at 1,000 v., and trunk mains, connecting Manchester Square, Sardinia Street and Rathbone Place, enabled one station to help another at times of heavy load. Neither the stations nor the individual machines were ever operated in parallel, but every alternator fed a group of circuits independently of the rest of the system.

PLATE X. SARDINIA STREET POWER STATION, 1896

PLATE XI. MANCHESTER SQUARE POWER STATION, 1890

The Manchester Square Station was able to start operations without any delay due to the laying of mains, because, by arrangement with the London Electric Supply Corporation, Ltd., it took over the overhead distribution system which that Company had established in the Marylebone area. This piece of good fortune, however, was more than neutralized by complaints, almost from the day that the station was started, as to the nuisance it caused to residents in the neighbourhood on account of the vibration it set up in their houses. Legal action was taken against the Metropolitan Company, and evidence was given that the vibration was so bad that it caused the stoppage of clocks on the walls, while the rattling which accompanied it was recorded on a phonograph by Professor S. P. Thompson for the edification of the Court. An injunction was granted against the Company, who spent large sums on improving the foundations of the machinery. All efforts, however, proved ineffectual, and by 1894 it seemed as if the station, which had cost £80,000, would have to be shut down entirely. The situation was saved at the last moment by the substitution of Parsons turbo-alternators for the reciprocating units that were causing the trouble. These turbo-alternators had an output of 350 K.W. at 1,000 V. 100 cycles and 3,000 R.P.M. They were notable as being not only the first machines of the kind to be installed in any power station in the Metropolis, but they were of more than twice the capacity of any turbo-generators that had been constructed up to that time.

A vivid account of experiences at Manchester Square, and by inference, a picture of the possibilities of central station life in the nineties, is to be found in the following extracts from letters by Mr C. A. Holbrow and Sir William A. Tritton which appeared in *The Engineer* during February and March 1934. Mr Holbrow wrote:

I was one of the engineers of the Manchester Square Station when the first turbines were installed there. The high-pressure portion was considerably the longest, and this perhaps was a good thing, as the second expansion blading often suffered from what we then learned was the "whipping" of the shaft. This did not worry us very much as it was a simple matter to cut out the damaged blading, close up the set, and continue working. Sometimes nearly one third of the total blading would be cut out, but the remainder was quite sufficient to help us through the time of heavy load.

In twelve or eighteen months after the first turbine was put on commercial load we had gradually eliminated most of the difficulties, and when the heavy load came on in the winter we found the advantage of using turbines as compared with the reciprocating engines. With the reciprocators, if a small quantity of water came over with the steam, it generally meant damage to the central valve trunks and distance pieces, which took some considerable time to repair and was very costly. With the turbine, when we had a partial strip, we generally ran on to the time when it was possible to disconnect the set from the supply, lift the cover, cut out the damaged blades, and have the set ready for running by the next evening's load. At this time we had a double shift of bladers, but it was so necessary to run the turbines rather than the reciprocating engines on account of the vibration caused by the latter, that the bladers could only get at their work at stated times and at the week-ends.

At times, owing to the turbine spindles and the rotating armatures of the alternators getting out of balance, we had some little bearing trouble, but with the concentric bush bearing and the tubes that acted as sleeves over it, it was a very rare thing for the spindles to seize. When they did, the shaft broke at a nick cut for that purpose in the square end which entered the coupling, and no further damage was done.

It was a most extraordinary thing how long it was possible to keep one of these turbines going when a hot bearing occurred. The driver used to cover the bearing up with cotton waste, and we had a cold water circulation which kept the waste wet. Perhaps after an hour or more, the bearing was still heating, with a quantity of smoke coming out. We then used heavy cylinder oil, and this either had the effect of cooling the bearing down, or else the smoke became so great that we could not live in the engine room.

The full-load steam consumption of the reciprocating engines was about 28–30 lb. per B.H.P.-hour. The statement of "Steam per K.W.H." was not much used in those days, possibly because it made the consumption appear higher. It must be remembered that the consumption figures for the engines referred to full-load conditions only, the half-load consumption being twice as great. We were told, of course, not to run the reciprocators at half load, but what could be done when, at four o'clock on a winter afternoon, we started with a head of steam of 160 lb.—10 lb. more than the normal pressure—and by eight o'clock the pressure was down to nearly 100 lb. per sq. in.? And this in spite of the fact that we had a forced draught fan capable of giving a pressure of nearly 4 in. w.g. under the grates, and at times the flames could be seen coming out of the top of the 120 ft. chimney.

It would be interesting to know what would be the leaving losses of the boilers under such conditions. It may be mentioned that the lining of the flues was burned out, and the supports that kept the exhaust pipe in the centre of the chimney were all found at the bottom in a confused mass. The exhaust pipe sagged to the side of the chimney, but did not break so that the Station could continue to be operated.

The turbines showed up to advantage when working with a falling steam pressure, as they would carry the load till the pressure fell to about 120 lb. After this point, when the steam relay ceased to function, it was a common practice to hold the relay valve full open with a cork, or by tying a weight on to the starting lever. By such means a continuous flow of steam passed through the turbine, and this enabled the latter to maintain the voltage under conditions when the reciprocating engines would have been useless, for want of steam pressure. As regards the efficiency of the turbines, it was said that they used about 60 lb. of steam per K.W.H., but this figure was probably on the high side. In those days the question of continuous running was much more important than that of steam consumption. I well remember that, after twelve months' working, the Company's auditors showed that the total cost of running the Manchester Square Station with turbines was less than it had been in previous years when the reciprocating engines only were in use.

The original alternators had surface-wound armatures with the conductors held down by piano wire binding. The armatures heated up considerably, and frequently the piano wire fractured and was forced out at the ends of the machine. At such times there was a lot of wire whipping round each end of the armature, and a good deal of fire flying about. After one or two burn-outs the men in charge got to know by the variation of the hum of the machine when a burn-out was likely to occur, and a man was then kept at the stop-valve ready to shut down. No one else was allowed to be near.

These surface-wound armatures were soon discarded and replaced by tunnel-wound armatures, but even these did not withstand the many short-circuits that occurred on the distribution system. During a heavy thunderstorm one or other of the sixty or seventy feeders going out from the station generally shorted, and the rule was to replace the fuse once before cutting out the defective feeder. This repeated the short-circuit on the armature. At other times it was necessary to burn out a cable which was only slightly defective, a practice which put a considerable overload on the machine. We had some five or six spare armatures, though at times there was only one or two of them available for service. The damaged ones were at first returned to the Heaton Works but we afterwards rewound them on site. The conductor was in a single length, so that by the time it had been drawn through all the tunnels, the insulation near the end had become considerably frayed. At one time we encouraged the winder by giving him a bonus if his armature would stand up for 24 hours, as this did not always happen. If we got over the first two or three days we expected the armature to last a month or two. But, like the difficulties with the turbine, after eighteen months or so all these frequent breakdowns came to an end, and we had a comparatively easy time until a short-circuit occurred at the back of the main switchboard. This burned out the switchboard and the engine room and boiler room roofs. Twenty-two fire engines were playing water over the turbines for about four hours, but by the following morning we had obtained from the Great Western and other Railway Companies sufficient tarpaulins to shelter the turbine-room, and as the steam pipes had not been damaged two of the Parsons sets were arranged to supply current to feeders directly connected to the alternator slip-rings, and were put into service again by this means the same evening. The remaining load was taken by the inter-connected stations.

Mr Holbrow's reminiscences were followed by a letter from Sir William Tritton, in the couse of which he said:

My most vivid recollections of Manchester Square are:

(a) The interminable trouble of balancing. The turbines were not too bad, but the armatures were awful. The critical speed was about half the running speed, and we had to warm up and then jerk the revolutions up as quickly as possible, otherwise the set started walking down the engine-room.

(b) Rewinding armatures. In one case this took over two years and we never got it to run decently. As the cable had some necessary clearance in the punched laminations, it was always loose and when under load it moved about, thereby upsetting all our efforts of balance. We had no balancing devices other than running the machines under steam.

(c) Hot bearings. These were cured (!) by shifting one or more of the pole pieces apart. This was done by inserting bits of key steel between one or two of the yokes. In some cases the magnetic field had $\frac{1}{4}$ in. air gaps in the iron circuit.

(d) Back pressure under full load up to 20 lb. in the exhaust main was registered.

(e) Shortage of boiler power. With forced draught we usually had a

PLATE XII. GENERATOR PANELS AT MANCHESTER SQUARE
POWER STATION, 1890

PLATE XIII. FEEDER PANELS AT MANCHESTER SQUARE
POWER STATION, 1890

plume of flame 20 ft. long on the top of our chimney shaft, much to the consternation of the local householders.

(*f*) Our bottom row of boiler tubes was generally bunged up solid.

(*g*) Our switchboard fire. This brought the roof down, and all but two of the sets were damaged and could not be run. The temporary switch-gear was made up of anything handy, and we had no ammeters or voltmeters in circuit. As the alternators could not then be paralleled, we had to change over by slackening the cable thimbles, and we broke the circuits under load by means of a snatch on a rope. We did not bother too much about putting all the lights out then in the West End several times a day in order to change over.

With reference to item (*d*), it should be explained that a back pressure was deliberately put on the Manchester Square turbines at times of low load by means of a valve in the exhaust pipe. The object of this was to minimize the effect of the pulsations in the exhaust due to the action of the "puff governing" which was then customary with Parsons turbines, as the pulsations caused complaints from neighbouring householders on account of windows rattling etc.

The fire mentioned above took place on 27 June 1898. The Amberley Road Station was about to be shut down, and the load taken over by Manchester Square. The operator there ran up a machine for the purpose, but, instead of connecting it to the Amberley Road trunk main, he inadvertently put it into parallel with another running machine, by making use of a wrong pair of plugs taken from another part of the switchboard. An arc was set up behind the board and the disastrous fire ensued. It was believed that an error of the same kind had been responsible for the fire at Sardinia Street in 1895.

The first of the 350 K.W. Parsons turbo-alternators was ordered in April 1894, and a few months later an order was given for two more. The next year three more were installed, the Willans engines which they replaced being transferred to Amberley Road, and other stations of the Company, where they performed useful service for many years. In 1900, in accordance with the intention of the Metropolitan Company to convert their distribution system from alternating to continuous current, the six 350 K.W. turbines in the Manchester Square Station had their alternators removed and replaced each by two Parsons direct-current generators coupled in tandem, to give 350 K.W. at 200–230 V. and 3,000 R.P.M.

The acquisition by the Metropolitan Company of rights to supply electricity in the Paddington district necessitated the building of a power station in that area. The site chosen was at St Peter's Wharf, Amberley Road, on the banks of the canal. The station, which was opened on 3 March 1893, was designed and constructed by the Metropolitan Company without the intervention of contractors. The boiler-house and coal store fronted on the canal, while the engine-room, offices and stores had access from Amberley Road. The engine-room contained five horizontal compound Hornsby engines with cylinders 15 and $24\frac{1}{8}$ in. in diameter by 27 in. stroke, running at 100 R.P.M. These were fitted with Meyer expansion gear controlled by Hartnell governors. Each engine drove a 120 K.W. 1,000 V. Oerlikon single-phase alternator at 600 R.P.M. by means of eight cotton ropes from a 14 ft. flywheel mounted on one end of the crankshaft. The drives were 25 ft. between centres. The engines ran non-condensing, which may appear curious in view of the proximity of the canal, but there was some difficulty about obtaining permission to make use of the water. The decision to adopt slow-speed engines with rope drives at so late a period is also somewhat surprising, but the action of the Company in this matter may have been influenced by the unfortunate experience they were having with vibration from direct-coupled sets at Manchester Square. There seems to have been no trouble with the machinery at Amberley Road, so far as vibration was concerned, but at first the humming of the alternators was loud enough to be objectionable. An attempt was made to remedy this by boarding up the spaces between the arms of the rotors, but this proved ineffective, and the noise was finally suppressed by fitting wooden packing between the armature coils.

As at the other stations of the Metropolitan Company, no provision was made for operating the alternators in parallel. The leads from the machines were taken to a marble switchboard on a gallery in the engine-room, which had arrangements for plugging

PLATE XIV. AMBERLEY ROAD POWER STATION, 1893

From contemporary engraving in *The Engineer*

the various feeder circuits on to the machines as required. It is recorded, however, that on the occasion of the official opening of the station, two of the alternators were connected in parallel for demonstration purposes, one motoring the other, the engine of which had its valve-chest covers and cylinder covers removed.

Steam was supplied to the engines from five Hornsby boilers of the locomotive type, working at 160 lb. pressure. These were replaced in a short time by Babcock and Wilcox boilers, the number being increased to eight, and the generating capacity of the station was increased by the erection of six Willans-E.C.C. single-phase 1,000 v. units of 150 K.W. each, taken out of the Manchester Square Station. In 1895 the Company arranged to take over an order which the City of London Electric Lighting Co. had placed with Messrs C. A. Parsons and Co. for four 500 K.W. 3,000 R.P.M. turbines, and had these machines fitted with two-phase 1,100 v., 100 cycle alternators and erected at Amberley Road. These sets were replaced in 1902 by four Parsons two-phase turbo-alternators of the same capacity but running at only 1,800 R.P.M. and generating 60 cycle current. All the turbines, like the reciprocating sets, were non-condensing, but one of the turbines was later equipped with a Korting ejector condenser, by means of which a vacuum of 15 in. could be obtained.

The Amberley Road Station ran until 1926, when the supply was given by the London Power Company. It was thus the last of the small generating stations of the Metropolitan Company to be shut down.

In 1897 it had become evident to the Metropolitan Company that a new generating station would have to be built, as their existing stations at Manchester Square, Sardinia Street, Amberley Road, Rathbone Place and Whitehall Court were all approaching the limit of their capacity. The directors therefore determined to construct a plant capable of extension to provide for the ultimate requirements of the Company, so that the work of generation could all be concentrated where it could be most economically carried on, and the five existing stations eventually shut down. A site of eight acres was acquired at Acton Lane, Willesden, on the bank of the Grand Junction Canal and also served by two railways. Being situated some five or six miles from the centre of the demand, current had to be transmitted at high tension. The Metropolitan Company were, at the time, supplying nothing but continuous current from their Sardinia Street Station, and were contemplating replacing the whole of their alternating-current distribution by continuous current. This was to have been provided by the use of rotary converters, and these necessitated either a two-phase or a three-phase system of transmission so that they might be more readily started. Two-phase transmission was recommended by Mr Frank Bailey, with the approval of Dr John Hopkinson the consulting engineer to the Company, as it appeared to permit of simpler cables and the balancing of the phases was easier. A two-phase transmission system at 10,000 v. 60 cycles, to carry the current from Willesden to the other stations of the Company was therefore decided on, each phase being carried by a separate two-core concentric cable having the outer conductor earthed at the power station.

The Willesden Station was designed to have ultimately an engine room 384 ft. long by 112 ft. wide, flanked on each side by a boiler-house 384 ft. long by 88 ft. wide, with two chimneys for each boiler-house. To begin with, the boiler-house was made 164 ft. long, and equipped with 16 Babcock and Wilcox boilers of 3,500 sq. ft. heating surface, set in pairs and supplying steam at 160 lb. pressure superheated to 480° F. Eight of the boilers were hand fired and the remainder provided with Vicars' stokers, coal being delivered from an overhead bunker. There were no economizers. A single chimney 165 ft. high by 13 ft. 3 in. internal diameter at the top served the boiler-house. The whole of the generating machinery was ordered from the Westinghouse Electric Co. of Pittsburg, U.S.A., as no British maker had experience of two-phase generators of the size required, and the engineers' strike of 1897 made it impossible get reasonable delivery of British engines.

Three Westinghouse engines and alternators formed the first installation of machinery. The engines were of the enclosed compound vertical marine type, with cylinders 36 and 55 in. diameter by 36 in. stroke, and ran at 120 R.P.M. They were larger than any engines previously constructed at Pittsburg. They were fitted with piston valves, and developed 2,500 I.H.P. at 120 R.P.M. when supplied with steam at 120 lb. pressure and exhausting to atmosphere. With the normal steam pressure of 140 lb. (saturated) their average consumption between full load and half load was guaranteed not to exceed 22 lb. per I.H.P. or 34·69 lb. per K.W.H., assuming a combined efficiency of engine and alternator of 85% at full load. Behind each engine was a rectangular Wheeler condenser with a surface of 4,000 sq. ft. These were served by three Blake and Knowles vertical steam-driven twin air and circulating pumps, which, together with the boiler feed pumps, exhausted into a surface heater. The circulating water was taken from the canal, and cooled by three Barnard-Wheeler cooling towers, each about 12 ft. square and 37 ft. high. They were constructed of sheet steel with mats of woven wire inside, and a draught was provided by a 10 in. fan for each tower.

The Westinghouse alternators had each an output of 1,500 K.W. two-phase, 60 cycles, 500 v. at 120 R.P.M. Their armatures, 16 ft. diameter by 3 ft. wide, were mounted on an extension of the engine crankshafts which ran in bearings 14 in. diameter by 42 in. long. The end of each shaft carried a compound wound exciter giving 450 amp. at 100 v., which was sufficient for two of the alternators in case of emergency.

In a separate transformer house were 14 single-phase transformers of 250 K.W. each, for stepping up the current from 500 to 10,000 v. for transmission. These were constructed by the British Electric Transformer Co. They were not encased in tanks, nor had they any cooling arrangements beyond the natural circulation of air round them. The high tension circuits were controlled by air-break switches with arms 32 in. long, mounted high on the walls and operated by levers of the kind used in railway signal boxes. The 10,000 v. mains were switched on through charging gear similar to that introduced by Mr Gerald Partridge for the Deptford mains in 1893. The primary windings of the pair of step-up transformers having been connected to the bus-bars, the secondary windings were switched on to the mains through water resistances which were gradually cut out. The two phases were charged simultaneously, the charging gears being inter-locked. When charging was complete the resistances were short-circuited by fuses, and then disconnected from the circuit. Voltage regulation was effected by changing the tappings of the step-down transformers at the receiving end.

The Willesden Station went into commission in 1899 with the three generating units mentioned above. Duplicates of these were added in 1901 and 1902, while in each of the two following years a 3,000 K.W. Sulzer set was put down, bringing the capacity of the station to 13,500 K.W. of reciprocating machinery in 1904. In that year the first turbine was installed, this being a 3,000 K.W. Willans-Parsons machine driving a two-phase, 60 cycle, 2,750 v. Dick Kerr alternator at 1,200 R.P.M. This was followed in 1913 and 1914 by two Parsons-Dick Kerr turbo-alternators, each of 4,000 K.W. capacity at 1,800 R.P.M. It was not until 1923 that three-phase 50 cycle generating machinery was introduced to Willesden, when two Parsons turbo-alternators, each of 10,000 K.W. capacity at 6,600 v. and 3,000 R.P.M. were put down in the station. After that the amount of two-phase current sent out decreased rapidly, and the station was soon changed over almost entirely to the standard system.

An interesting feature in connection with the Willesden Station was the introduction there, by Mr J. S. Highfield, of the system of generation and transmission of power by means of high-tension continuous current, originated by Monsieur Thury. Three motor-generator sets were installed, each designed to produce a constant current of 100 amp. at a maximum pressure of 5,000 v. The generators worked in series, and supplied current to the Ironbridge substation at Hanwell, a distance of 5·5 miles. The system was in successful use for a number of years, the trans-mission pressure being finally raised to about 18,000 v. before the plant was taken out in 1924.

As regards the boiler plant, the original sixteen 10,000 lb. Babcock and Wilcox boilers were followed between 1900 and 1904 by an equal number of Fraser boilers of the Scotch marine type and of the same capacity and pressure. Babcock boilers were then reverted to, and in 1907 a start was made with 35,000 lb.

Babcock boilers, producing steam at 200 lb. pressure superheated to 500° F. A further advance was made in 1924 when Babcock boilers of the c.t.m. type with an evaporation of 50,000 lb. per hour, and a steam pressure and temperature of 275 lb. and 700° F. were adopted. It is worth recording that one of the very earliest experiments with pulverized coal for power station work was made in 1903 at Willesden. In that year a pulverized fuel apparatus was tried in conjunction with one of the 10,000 lb. Scotch boilers, but the guarantees of the makers as to efficiency were not fulfilled and the plant had to be removed.

The London Power Company took over the control of the Willesden Station from the Metropolitan Company on 1 January 1927. The last of the reciprocating sets, and other obsolete equipment, were then removed, larger boilers and turbines were installed, and the station thoroughly modernized.

THE KENSINGTON COURT ELECTRIC LIGHT CO.

Among the Companies which found a means of giving a public supply of electricity without having recourse to Parliamentary powers, and which, therefore, were able to carry on their businesses without regard to the prohibitive conditions imposed by the Electric Lighting Act of 1882, was the Kensington Court Electric Light Co. Registered in 1886 with a capital of £10,000, it had as its primary object the supply of electricity to the residents in Kensington Court, which occupied the land at one time forming the private estate of Baron Grant, a notorious Company promoter, whose unfortunate financial transactions led to the seizure of his property by his creditors. The Kensington Court Company obtained permission to lay their cables in the subways which existed under all the roadways, having been constructed for the purpose of accommodating hydraulic power mains. These subways, which were 6 ft. high and 3 ft. wide, passed round the entire estate, and gave easy access to the various buildings.

The whole of the Kensington Court undertaking was designed and carried out by Messrs R. E. Crompton and Co., who constructed and equipped the power station, laid the mains, and contracted to operate and maintain the plant for a period of three years. The power station was situated just off High Street, Kensington, and close to the subway in which the cables were carried. The plant started operations in January 1887, and was therefore one of the very earliest to be working in the London area. The machinery was at first housed temporarily in a wooden shed, and consisted of a Marshall boiler of the locomotive type and a Willans engine of the old launch pattern. The engine was lent by the makers pending the completion of a central-valve engine, which was late in delivery. The latter engine, when installed, was noteworthy as marking the beginning of the long supremacy of the Willans high-speed engine in central station work. It was coupled directly to a Crompton dynamo of 35 k.w. capacity, which it drove at 500 r.p.m., steam being supplied at 160 lb. pressure by the Marshall boiler. The flue was taken underground to a chimney outside the building. The Willans engine was of the single-crank compound design, and the Crompton dynamo to which it was coupled was of the very early type with double field magnets arranged horizontally, one pair on each side of the armature. It generated direct current at 100/140 v. Before the end of 1887 a second 35 k.w. generating set had been put down. This was also a Willans-Crompton unit of 35 k.w., but the dynamo was of the more recent type with vertical horseshoe field magnets.

The permanent station at Kensington Court consisted of a basement 88 ft. long by 50 ft. wide and 9 ft. below street level. This was occupied by the engines and boilers. In 1890 it contained three Babcock and Wilcox boilers of 5,000 lb. evaporative capacity, three 50 k.w. 450 r.p.m. Willans-Crompton sets generating at 100/120 v., and four 100 k.w. sets of the same make generating at 200/240 v., the two 35 k.w. units having then been removed to Cheval Place. The engines ran noncondensing, the exhaust steam being taken to a tank in which it imparted heat to the feed water, the excess steam escaping to atmosphere by a pipe reaching to within 20 ft. of the top of the 100 ft. brick chimney which served the station. The tank acted as a hot well from which the boilers were supplied by a pair of Worthington pumps. The main steam piping was of copper throughout, 9 in. in diameter over the boilers and 5 in. in the engine-room. To soften the feed water an apparatus designed by Messrs Babcock and Wilcox was installed. This consisted of a number of 6 in. pipes having 2 in. pipes fixed concentrically inside them. The water, on its way to the boiler, flowed in one direction through the smaller pipe and then returned through the annulus between the two. The outer pipe was subjected to the heat of the flue gases so that the apparatus acted as an economizer. The sediment and scale deposited in the tubes was blown out at intervals.

The supply to the consumers was given from a battery, with which, however, the generating set worked in parallel at times of heavy load. The battery consisted of 53 cells, each containing 35 plates 8 in. square. The plates were "formed" on the Planté principle, but instead of being plain plates of the usual Planté type, they were sawn out of ingots of lead cast, in such a way as to be porous, by a patent process of Messrs J. C. Howell of Llanelly. When fully formed, the cells had a capacity of 5 amp.-hours per pound of lead, or a total capacity of 600 amp.-hours per cell. They could be discharged without damage, in emergency, at the rate of 300 amp., which would empty them in 2 hours, and it was claimed that this rate could be considerably exceeded if necessary.

Nine steps of voltage regulation were obtainable by a hand-operated multiple contact switch controlling the regulating cells of the battery. The contacts were a series of brass rings, cut on one side so as to give a certain elasticity, and through them was forced a metallic brush by means of a screw and handwheel. When the dynamo and battery were working together on the load, the lights were connected across either 41, 42 or 43 cells, according to the state of charge. When the battery was working alone there were either, 50, 51, 52 or 53 cells supplying the load.

The method of operating the station was to start charging the battery before dusk. For a short time the dynamo would also supply the lighting load as well, but as the load grew heavier the battery would take its share. By about eleven o'clock at night the engine could be shut down, leaving the battery alone on the load.

The success of the scheme led to the extension of the mains beyond the limits of Kensington Court. They had been carried in the subways of the Kensington Court Estate, on wooden shelves fixed to the walls of the subways, but when they had to be taken farther afield, some system of laying them in the streets had to be devised. In some cases ordinary cables were used, drawn through 6 in. cast-iron pipes, or laid in bitumen, but the system most favoured by Colonel Crompton was to employ conductors of bare copper strip laid underneath the footways in brick culverts with concrete bottoms. The culverts were divided into two channels by a central partition of brick, and were covered directly by the flagstones of the pavement. The copper conductors were carried, one in each channel, on porcelain insulators supported on short uprights from the floors of the channels. They were strained in position by screw shackles, in order to reduce the number of insulators necessary and to prevent sagging. The system was cheap, and gave no trouble with the low voltage at which the main worked.

The lamps on the system were rated at 102 v., and gave from 10 to 12 c.p. for a consumption of 33 w. This size of lamp was generally used for domestic and shop lighting, although in some situations 66 w. lamps giving double the light were employed. Current was metered to the consumers by Ferranti or Cauderay meters, and was charged for at a flat rate of 8d. per K.W.H.

THE KENSINGTON AND KNIGHTSBRIDGE CO.

In March 1888 the Kensington and Knightsbridge Electric Lighting Co., Ltd., was registered with a capital of £250,000 to take over the Kensington Court Company and extend its business, and the next year it obtained a Provisional Order for supplying electricity in the districts from which it took its name. This enterprise called for a second generating station, which was established in Cheval Place in 1890, and a battery substation was put down in Queen's Terrace Mews. The new power station was equipped on the same lines as that at Kensington Court, namely, with Willans engines, Crompton dynamos and Howell batteries.

PLATE XV. BOILER HOUSE OF CHEVAL PLACE POWER STATION, 1889

In 1892 there were four Babcock and Wilcox boilers and seven generating sets of an aggregate capacity of 645 K.W. at Kensington Court, and three boilers with three generating sets aggregating 410 K.W. at Cheval Place. During the year 1891, one-eighth of the whole output of the Kensington Court Station passed through accumulators, which were said to have given an efficiency of over 85 % on a K.W.H. basis. For the whole system the K.W.H. efficiency averaged 79·5 %, this figure taking into account the losses in the charging mains to the substation. It is interesting to note that it was the opinion of Colonel Crompton that the amount of storage capacity advisable for low-tension continuous-current systems, such as that in question, should only be sufficient to supply one-fifth of the demand during a 24 hour December day.

Like so many of the early power stations, Kensington Court was the scene of a serious fire. In the evening of 12 February 1895, fire broke out in the accumulator-room, which was situated above the engine-room and fortunately separated from it by a concrete floor. The whole of the upper part of the building was burnt out, the batteries of course were destroyed, and matters were made worse by the ignition of the oil fuel for the boilers, though happily the main tank did not get alight. The damage done was estimated at £6,000, but the interruption of the current only lasted for about three hours, as the load was picked up by the station at Cheval Place.

By 1898 the load on the stations at Kensington Court and Cheval Place was something like 1,000 K.W., and it was evident that further generating plant would be needed to cope with the growth of the undertaking. To have extended either of the existing stations would not only have been difficult on account of lack of space, but their location in a residential area made it inadvisable to increase their capacity on account of the risk of creating a nuisance. The Notting Hill Electric Lighting Co., who were also operating a low-tension direct-current battery system from their station at Bulmer Place, established in 1891, found themselves in much the same position as the Kensington Company. The result was that the two Companies decided on the construction of a joint station at Wood Lane, Shepherd's Bush, in which high-tension three-phase current would be generated for transmission to substations belonging to the two Companies respectively. In these it would be converted to low-tension direct current by means of motor generators. The original scheme involved generation and transmission at 6,600 v., but when this was submitted to the Board of Trade, the voltage was considered too high for safety, although single-phase current at 10,000 v. had been sent out from Deptford for a number of years. The highest pressure that would be sanctioned for the Wood Lane scheme was 5,000 v., so this had to be adopted, and it remained the voltage of the Company until 1937 when the pressure was raised to 6,600 v., to be in accordance with that of neighbouring supplies.

The Wood Lane Power Station started operation in October 1900, and its inauguration is of historical importance because it provided the earliest example of high-voltage three-phase generation and transmission in this country, although it was destined soon to be overshadowed by the larger 11,000 v., three-phase station of the Charing Cross Company at Bow. Its first equipment consisted of three 5,000 v., three-phase 45 cycle alternators of 330, 550, and 750 K.W. capacity respectively, driven directly by Willans three-crank triple expansion engines. Steam was supplied by Babcock and Wilcox boilers. The 5,000 v. switches were of the air-break type, each fitted with a pair of horns up which the arc ran until its length became so great that it extinguished itself. The machinery was increased by the addition of a Parsons turbo-alternator of 1,000 K.W. capacity in 1903, and this completed the equipment of the first part of the station. Reciprocating machinery was reverted to for the second part, which was started in 1906 with a 1,600 K.W. slow-speed horizontal cross-compound Sulzer engine carrying a flywheel alternator on the centre of the crankshaft. The installation of this set afforded an opportunity of gradually raising the frequency of the whole system to 50 cycles. The superiority of the turbine over the reciprocating engine, however, soon reasserted itself, and all subsequent extensions were made with turbo-generating plant of one kind or another. In 1916 a beginning was made with the elimination of the reciprocating machinery, and when the Wood Lane Station was shut down in 1928, turbo-generators alone remained, their aggregate capacity being 13,700 k.w.

Returning to the system of the Kensington and Knightsbridge Company, the original supply pressure of 100 v. was raised to 200 v. when suitable lamps had become available, and in 1903 a beginning was made to convert the 200 v. two-wire network to a three-wire system with 400 v. across the outers, a change that required three or four years for its completion. The high-voltage supply obtained from Wood Lane in 1900 was dealt with by means of motor generators installed in a substation in the Albert Vaults, close to the Albert Hall, and in the Kensington Court Station, thus allowing the steam plant in the latter station to be shut down. In 1923 an additional source of high-voltage supply was made available by connecting the Cheval Place Station with the Grove Road Station of the Central Electric Supply Co. This current was brought in by 6,600 v. feeders, and the pressure reduced by auto-transformers to 5,000 v. Cables working at the reduced pressure enabled a supply from Grove Road to be given also to Kensington Court and the Albert Vaults substation, via Cheval Place.

From 1922 onwards, a few large consumers, including the South Kensington Museums and Colleges, were given alternating-current supplies at high voltage, which they transformed down to low-tension current for themselves. The ordinary consumers continued to receive a direct-current service until 1929, when a start was made to convert the whole of the distribution system from direct to alternating current.

THE NOTTING HILL ELECTRIC LIGHTING CO., LTD.

This Company, which was incorporated on 21 February 1888 with a capital of £100,000 to provide a service of electricity in the Notting Hill district, constructed its power station at Bulmer Place. The system of supply was similar to that of the Kensington and Knightsbridge Electric Lighting Co., namely, low-tension direct current distributed by a three-wire network. The Bulmer Place Station was formally opened by Sir William Crookes,

F.R.S., on 1 June 1891. It started with three Willans-Crompton generating sets, the largest having a capacity 120 K.W. at 240 V. and 350 R.P.M., and the two smaller ones being each rated at 50 K.W. at 120 V. and 480 R.P.M. The large dynamo, which was a four-pole machine, was connected across the outers of the three-wire system, while the smaller two-pole machines each supplied one side of the network. In order to avoid troubles from vibration, the engines and dynamos were erected on a solid concrete raft 10 ft. thick, supported on piles, and of sufficient area to accommodate five more 120 K.W. units.

Steam was supplied at 140 lb. pressure by two Babcock and Wilcox boilers rated at 350 H.P. each, and two Marshall boilers of the locomotive type, rated at 75 H.P. each. The latter were removed after a few months, leaving room for three more Babcock and Wilcox boilers. Two 500 amp.-hour Crompton-Howell batteries, each consisting of 56 cells of 61 plates, were also installed.

The engine-room and boiler-room were both in a basement, the floor level of which was 22 ft. below the street level. Coal could be wheeled to the front of the boilers from an underground coal store, but the barrows had to pass through the engine-room on the way, and the position of the boilers made it difficult to fire them without scattering coal among the machinery. The layout of the station was, therefore, hardly one that could be commended, and it is not surprising that after a few years' working, the Notting Hill Company found it to their advantage to join forces with the Kensington and Knightsbridge Company for the construction of a joint station at Wood Lane, from which both undertakings could obtain a supply in bulk. This station, which is referred to on p. 94, went into service in October 1900.

The switch-gear at Bulmer Place was supplied by Messrs Crompton, and was a duplicate of that used in the Kensington Court and Cheval Place Stations of the Kensington and Knightsbridge Company. The mains were also of the Crompton type, that is to say, bare copper strip stretched between insulators and laid in culverts under the pavements. Where this construction was impossible owing to the presence of cellars, etc., the conductors took the form of cables drawn into iron pipes.

THE WESTMINSTER ELECTRIC SUPPLY
CORPORATION, LTD.

The first public supply of electricity in Westminster was provided by the Westminster Electrical Syndicate, which operated under a Board of Trade Licence granted to it in 1888 for a period of seven years. The Syndicate furnished a direct-current supply at 100 v. by means of overhead lines. It possessed two small generating plants, one at the Stoneyard, Millbank, near the House of Lords, and the other in Chapel Mews not far from St James' Park Station. The former installation consisted of two Marshall semi-portable steam-engines driving Goolden dynamos of 32 and 40 k.w. respectively, its principal duty being the lighting of the Houses of Parliament, for which the Syndicate had obtained the contract in 1889. The equipment of the other plant was very similar, namely, dynamos driven by belts from semi-portable engines.

In June 1888, the Westminster Electric Supply Corporation was registered, with a capital of £100,000 (increased the next year to £300,000), to develop the supply of electricity in the area on a larger scale. It applied for the requisite Provisional Orders, and as a result of the Board of Trade inquiry in 1889, valuable districts in Westminster and Mayfair were allotted to the Company. It was not, however, granted a monopoly, for it had to compete both with the Westminster Electrical Syndicate and with the London Electric Supply Corporation, because, in accordance with the Board of Trade policy of allowing consumers a choice between alternating and direct current, the latter Company was also empowered to give a service in the same area. Competition with the Syndicate was terminated by the purchase, in January 1890, of the undertaking and assets of the Syndicate for £10,000. Furthermore, the burning out of the Grosvenor Gallery Station in November of the same year, and the troubles experienced by the London Electric Supply Corporation in connection with its other station at Deptford, not only crippled that Company for many months as a competitor, but resulted in a large number of its customers transferring their allegiance to the

Westminster Company. After some years the London Company ceased to compete at all, and it ultimately gave up its rights in the area.

The Westminster Company gave its first supply in 1890 from a temporary station in a yard off Dacre Street, behind Victoria Mansions. When it took over the business of the Syndicate, among the property acquired was the lease of an admirable site for a power house to supply the Westminster neighbourhood at St John's Wharf, Millbank, where coal could be brought by barge and where ample water for condensing purposes was obtainable from the Thames. Other sites were purchased in Eccleston Place and Davies Street, from which the needs of Belgravia and Mayfair respectively could be conveniently supplied. Current was first sent out from the Millbank Station in November 1890, from Eccleston Place Station in February 1891, and from the Davies Street Station in March 1891. The three stations were designed by Professor A. B. W. Kennedy who was chief engineer of the Westminster Company from 1890 to 1926, and represented the most advanced practice at the time, so far as the low-tension direct-current system of supply was concerned. They all worked in parallel, delivering current to a three-wire network with 220 v. between the outers. Storage batteries were provided at all stations, both as a safeguard against interruptions of supply and to assist in maintaining a steady voltage, but perhaps their most important function was to enable the engines and boilers to be shut down entirely at times of light load, and operated at their most economical ratings when in service.

The Eccleston Place Station may be regarded as typical of the engineering practice of the Westminster Company in the early nineties. The station buildings were of brick, the boiler-house and the engine-room being alongside each other on the ground floor, and the battery-room above the boiler-house. An unusual refinement for those days was the provision of a mess-room for the comfort of the operating staff. The boiler-house contained four internally fired boilers, one by Messrs Fraser and Fraser, and the other three by Messrs Davey, Paxman and Co. The latter were of the firm's "Economic" type, each having two furnaces in cylin-

PLATE XVI. ECCLESTON PLACE POWER STATION, 1892

From contemporary engraving in *The Electrical Review*

drical flues, and a large number of small fire-tubes through which the gases returned to a smoke box at the front of the boiler. The design was generally similar to that of a Scotch Marine boiler, except that there were no internal combustion chambers. The boilers generated steam at 150 lb. pressure and delivered it in the saturated condition to the engines. The feed water was pumped through a Kennedy positive water meter and then through a pair of Green's economizers before entering the boilers. The economizers, which comprised 400 cast-iron tubes, raised the temperature of the feed to about 160° F. There was no other provision for feed-heating, the economizer having to deal with cold raw water, as the engines worked non-condensing. Steam was distributed by a steel ring main with all bends and branches to the engines of copper and stop valves of cast iron. Ring mains were much favoured by Professor Kennedy, and one of cast iron was provided to collect the exhausts of the engines before they passed into the chimney.

The generating machinery consisted, in 1892, of six direct-coupled engine-driven dynamos with an aggregate capacity of about 450 k.w. The station started with a pair of 33 k.w. 110 v. Siemens dynamos, each driven by a compound Willans engine having cylinders 10 and 14 in. diameter by 6 in. stroke. These were soon followed by a pair of 67·5 k.w. units, and a pair of 120 k.w. units. The dynamos of these four sets were all four-pole Crompton machines, the two smaller ones being driven by vertical open-type two-crank compound engines by Davey, Paxman and Co., and the others by two-crank triple-expansion Willans engines with cylinders of 10, 14 and (two) 20 in. diameter by 9 in. stroke, capable of developing 200 i.h.p. at 355 r.p.m. These engines were designed to work with jet condensers which were fitted later. The smaller Crompton dynamos generated at 225 v. and the larger ones at 240 v. These machines took care of the heavy load, the Willans-Siemens sets being used mainly for the day supply and for battery charging, being also employed on the evening shift for balancing the load.

The storage battery consisted of 56 Crompton-Howell cells standing on wooden racks, each cell having a capacity of 500 amp.

hours. The battery was charged in parallel with the load during the daytime when the demand for current was small. By the evening, charging was complete and the cells were switched off until the load had fallen to an amount which the battery could safely carry. The engines were then stopped and the station shut down, the supply being maintained by the battery alone until the next morning. The operators endeavoured to ensure that the full nominal capacity of the battery was charged and discharged once every 24 hours to keep the cells in good condition. The control of the battery was effected by the aid of a modified Aron meter in which the usual permanent magnets had been replaced by electro-magnets. This meter was set at zero when discharge commenced, and recorded the number of ampere-hours discharged. In due course recharging was started, and this was continued until the reading of the meter was once more zero, showing that the amount of current put back had been equal to the amount taken out of the cells.

A flat-back switchboard of slate, supplied by Messrs Crompton and Co., served for the control of the generator and feeder circuits. Panels were arranged for 14 dynamos and 8 feeders. The board was erected on a raised gallery along the engine-room wall with stairs for access at each end. Copper bus-bars were mounted on the back of the board, which stood out 2 ft. from the wall so as to render the connections accessible. Every generator circuit was provided with a reverse-current automatic cut-out on the negative side to prevent the dynamo being motored by the battery or by other machines. The leads from the dynamos to the board were drawn into 3 in. gas pipes concealed in the flooring and walls of the engine-room. The feeders outside the station consisted in general of bare copper strip supported on insulators in culverts underneath the pavement. Where there was no room for culverts rubber-insulated cables were drawn into iron pipes or earthenware ducts.

The first supply of condensing water for the station was ob-tained from a large sump which collected the rain-water from the roofs of Victoria Railway Station. To get this water to Eccleston Place it was necessary to lay a pipe along Buckingham Palace

Road, which was done under the Statutory Powers possessed by the Company to open streets for the purpose of laying "mains". This interpretation of the word did not altogether please the Local Authorities when they discovered what the excavations had really been made for, but there was apparently nothing to be done about it, and the pipe remains to this day, serving now as a duct for cables. As the demand for cooling water increased with the growth of the plant, an additional supply was obtained by laying two 16 in. pipes to the Grosvenor Canal, some 500 yards distant from the station.

By 1889 the Westminster Company possessed 41 Willans engines of an aggregate capacity of 6,100 K.W. in its three stations. The load on the system had attained 5,000 K.W., and as none of the existing stations could be economically extended, the Company made arrangements with the St James' and Pall Mall Electric Lighting Co., which had found itself in much the same position, for the construction of a new joint generating station at Grove Road, Marylebone. The Grove Road Station, which went into service in November 1902, furnished three-phase current at 6,600 v. to the three stations of the Westminster Company, and to the Carnaby Street Station of the St James' Company, where it was converted to direct current by motor generators. In 1904 a substation with four 400 K.W. motor generators was put in commission at Duke Street, Mayfair. By 1928 this substation contained five 2,000 K.W. and three 1,500 K.W. rotary converters, or a total of 14,500 K.W. of running machinery, but this has since been almost entirely replaced by static transformers taking current from the 66,000 K.W. mains of the London Power Company.

The first of the generating stations of the Westminster Company to be closed was that at Millbank, which was demolished in 1910 when the site was wanted for the construction of the Victoria Gardens. It was replaced by a new station built in the Horseferry Road, near the end of Lambeth Bridge. The first equipment consisted of three Howden turbines, each driving a pair of 500 K.W. direct-current generators in tandem at 1,850 R.P.M. There were also two balancing sets of similar type, each having a pair of

150 k.w. dynamos. In 1913 a 1,500 k.w. Parsons impulse-reaction turbine, driving a 410/460 v. Siemens dynamo at 300 r.p.m. by means of 10 : 1 reduction gearing, was installed, and this was followed in 1919 by a cross-compound Parsons turbine of the pure reaction type, driving a 3,000 k.w. Siemens dynamo with a similar speed reduction ratio. Both sets were provided with surface condensers, giving vacua of 28·5 and 29·0 in. respectively. The turbines worked with steam at 200 lb. pressure superheated to 538° F., supplied by seven Babcock and Wilcox boilers, each rated at 20,000 lb. evaporation per hour. As a consequence of the co-ordination of electrical generation by the London Power Company, the steam plant at Horseferry Road was scrapped in 1928, and replaced by four 3,000 k.w. rotary converters, which were the largest machines of the kind in the country at the time.

The steam plant at the Davies Street Station was finally shut down in 1921, and that at Eccleston Place in 1922. The engines at Davies Street had always been non-condensing, as no supply of water for condensing could be obtained. In spite of this handicap, the Company, in its early years, operated ice-making machines in this station and also at Eccleston Road during the summer months in order to provide a useful load when the demand for electricity was small.

In course of time the limitations of the direct-current system of supply began to make themselves evident, and in 1925 the Westminster Company experimented with a six-phase alternating-current system, and supplied several large consumers in this manner. It was, however, soon decided to adopt the standard three-phase system. Before the end of 1927 a supply on that system was available in the neighbourhood of Piccadilly, and now little remains of the original direct-current service.

THE ST JAMES' AND PALL MALL ELECTRIC LIGHTING CO.

The St James' and Pall Mall Electric Lighting Co., Ltd., was registered in March 1888, and commenced supply on 4 April 1889 from a power station in Mason's Yard, Duke Street. This was in the heart of Clubland, the station being placed almost centrally in the rectangle bounded by Piccadilly, Pall Mall, St James' Street and Lower Regent Street. The Company put up a fine new three-storey building for offices and workshops, but the operative part of the station was entirely underground, the engine-room and boiler-room being in excavations beneath the yard. The first machinery consisted of a pair of 80 I.H.P. Willans engines driving 50 K.W. Latimer Clark dynamos at 475 R.P.M. and four 210 I.H.P. Willans engines driving 120 K.W. Siemens and Latimer Clark dynamos at 340 R.P.M. The station was completed by 1891, when it contained two 80 K.W. sets and ten 120 K.W. sets, all of the type mentioned. These were arranged in two rows on either side of a central gangway. They were supplied with steam at 150 lb. pressure by six Davey-Paxman boilers of the locomotive type, each with a tube surface of 1,593 sq. ft., a fire-box surface of 146·8 sq. ft., and a grate area of 32·5 sq. ft. Their evaporative capacity was 8,000 lb. per hour, which was considered very large at the time. Feed water was obtained from an artesian well sunk by the Company on the premises.

The engines worked non-condensing. Their exhausts were led by cast-iron pipes beneath the engine-room floor into two Berry-man feed-water heaters, the excess exhaust steam being discharged into the chimney. This chimney was a cause of offence to the dignified inhabitants of the neighbourhood, who complained of it even before it was put into service, and much more afterwards. There was said to be quite a good business done in top hats by the Piccadilly shops, owing to the oily spray discharged with the exhaust steam from Mason's Yard, and in the words of one of the engineers who was there at the time "it was almost a daily occurrence to see a gesticulating man pointing out a damaged topper to one of the Company's staff, the complainant, if outside the station, being tactfully interviewed on the windward side".

The Company was also subjected to lawsuits on account of vibration, and one of the first consumers, the Wyndham Club in St James' Square, obtained an injunction against the undertaking, but nevertheless did not want their supply cut off! As a compromise, the operation of the injunction was deferred for three months. Meanwhile it had become evident that the two-crank Willans engines were the cause of the trouble, and as at that time Messrs Willans and Robinson had undertaken the construction of three-crank engines up to about 200 K.W., one of this type was at once installed. The result was so satisfactory that an order was placed for a full equipment of similar plant. Work on the Carnaby Street Station, which was then in course of construction, was also accelerated as much as possible, in order to be able to relieve Mason's Yard in case of necessity, and a few days before the injunction would have become operative the old two-crank plant could be dispensed with. In connection with the vibration difficulties, it is interesting to note that a portable recording vibrometer was designed by Mr Stanley Peach—no such instrument being then on the market—and this was wholly constructed by the staff of the St James' Company. The instrument, though crude in comparison with modern apparatus of the kind, worked most successfully and was employed for many years in the investigation of cases of alleged vibration.

Another difficulty arose in connection with the discharge of hot water from Mason's Yard into the sewers. The Company was prosecuted by the Attorney-General on behalf of the Local Authorities for allowing hot water to enter the sewers in such quantities as to render it impossible for the sewer men to carry out their work. This action, like the others, was settled amicably, as means of overcoming the objection were found. The Mason's Yard Station continued in service as a generating station until 1910, when it was employed as a substation to deal with a bulk supply of current from Grove Road.

The dynamos were all shunt-wound undertype bi-polar machines, designed to generate at 120 v. The supply at first was given by means of V.I.R. cables run overhead "among the chimney pots" to buildings in the vicinity. This system was replaced as soon as possible by underground mains. The dynamos delivered current into a three-wire network with 214 v. across the outers, the machines running in pairs across the system. The mains consisted of bare copper strip, 2 by 0·1 in. in cross-section, eight of such strips placed side by side being used for the largest conductors. They were laid in cast-iron culverts, in which they were spaced and supported by slots in porcelain bridge pieces. The culverts were closed by cast-iron lids bolted on. Some of these mains were in use as late as 1928.

Leakage of gas into the culverts resulted in more than one explosion. The first happened in a street box at the corner of Duke and Jermyn Streets, the cover of the box being deposited on the slate roof of the annex to a near-by hostelry, whereupon the proprietor withdrew the hospitality which he, like so many other consumers in the neighbourhood, had been wont to lavish on the staff of the Company. The explosion resulted in a Board of Trade inquiry, as did another in Jermyn Street a little later, and indeed such inquiries formed quite a feature of the Company's existence during a certain period. A street-box accident of a different kind occurred one day when, owing to the cover of the box becoming alive, an old horse met with a fatal shock opposite St James' Hall. The Company paid £40 as compensation to the owner, and the sequel was rather extraordinary. Very shortly afterwards another horse, drawing a hansom cab, was killed in a similar manner, which cost the Company £60 and a Board of Trade inquiry. It was then discovered that some of the local cab-drivers had organized a regular patrol of old and worthless horses up and down the street, in the hope of further accidents of the same kind! Horses, it may be said, are particularly sensitive to electric shock from stepping on live metal, no doubt on account of the good contact made by their shoes, and a number of fatalities are on record from this cause.

The fire at Grosvenor Gallery Station on 15 November 1890 and the consequent cessation of supply by the London Electric Supply Corporation for a period of three months, resulted in a large number of the customers of that Company turning to the St James' Company, who were not slow to take advantage of the situation. Partly on account of this unexpected piece of good fortune, the demands on the Mason's Yard Station were approaching the limits of its capacity in 1891, so the Company decided on the construction of a new and larger station in Carnaby Street.

This latter station was opened in 1893 with five Willans-Siemens direct-current units of an aggregate capacity of 650 K.W., taking steam from six Davey-Paxman "Economic" boilers of the dry-back marine type. The equipment had almost continually to be increased, and by 1900 the capacity of the plant installed had reached 4,480 K.W. The system of supply was the same as that at Mason's Yard, namely, three-wire direct current at 214/107 v. As at the older station, the whole of the engines ran non-condensing, and water for the boilers was obtained from an artesian well of 10 in. bore and about 450 ft. deep. The normal output was 10,000 gallons per hour, and the water was lifted by an electrically driven reciprocating pump, the pump and gear being constructed by the staff of the Company. When the Carnaby Street load was transferred to the Grove Road Station, the bore-hole was sealed up, but before this was done somebody threw a number of current French coins down it, in the hope of puzzling some future archaeologist.

The last of the generating plant at Carnaby Street was shut down in 1923, when the station was reconstructed to serve as a substation for the distribution of a bulk supply of current from Grove Road.

THE CHARING CROSS ELECTRICITY SUPPLY CO., LTD.

The beginnings of the Charing Cross Company can be traced back to a private installation put down in September 1883 by Messrs A. and S. Gatti to light their Adelaide Restaurant in the Strand. The plant, erected in the basement of the building, consisted of a pair of vertical multi-tubular Field boilers supplying steam to two Armington and Sims engines, each of which drove a pair of 150-light Edison dynamos by means of belts and shafting. The lighting of the restaurant was done by 330 incandescent lamps. In 1885 mains were laid to give a supply to the Adelphi Theatre, and in 1888 Messrs Gatti established a new generating station in Bull Inn Court, between Maiden Lane and the Strand. This station, which was below ground level, was equipped with three Babcock and Wilcox boilers working at 140 lb. pressure, and

PLATE XVII. MAIDEN LANE POWER STATION, 1889

From contemporary engraving in *The Engineer*

four compound Willans engines direct coupled to Edison-Hopkinson dynamos. Two of the generating units had a capacity of 84 K.W. at 425 R.P.M., and the other two a capacity of 50 K.W. at 475 R.P.M., all producing continuous current at 105 V. The dynamos were not worked in parallel, but èach supplied current to its own pair of bus-bars to which any desired number of the twelve lighting circuits could be connected by plugs. To avoid sparking when changing circuits from one dynamo to another, the plugs could be temporarily short-circuited by a quick-break switch. To change over the generators, the incoming machine was run up to load on a resistance, and the change–over was then effected by the reversal of a commutating switch.

The Maiden Lane Station was taken over from Messrs Gatti in 1889 by a Company called the Electric Supply Corporation, Ltd., registered in that year with a capital of £100,000. This Company obtained a Provisional Order in 1889 for supplying electricity in the Parish of St Martin-in-the-Fields, in competition with the Metropolitan Company who supplied alternating current, and changed its name to the Charing Cross and Strand Electricity Supply Corporation, Ltd. They reorganized the Maiden Lane Station, making the engine-room on the ground floor and using the basement for the boilers. A second chimney stack was built, and the boiler plant increased to seven units, all of the Babcock and Wilcox make. A well, 184 ft. 6 in. in depth, was sunk in the boiler-room for the supply of feed water. In 1892 the Maiden Lane Station contained five Willans engines of 200 I.H.P. at 350 R.P.M., three of 125 I.H.P. at 375 R.P.M., and two of 80 I.H.P. at 475 R.P.M., all directly coupled to Edison-Hopkinson or Siemens direct-current generators working at 115 V. and with an aggregate capacity of about 900 K.W. The Company also changed the distribution system from two-wire at 100 V. to three-wire at 200/100 V. soon after taking over the undertaking.

In 1893, Mr W. H. Patchell, the engineer to the Company, made some of the earliest experiments with superheated steam for power station work. The superheater he employed was of the McPhail and Simpson type, consisting of a bank of hair-pin tubes arranged beneath the drum of one of his Babcock and Wilcox boilers. From these tubes the steam was taken through another set of tubes immersed in the water in the boiler drum. These latter tubes, of course, eliminated almost the whole of the super-heat. The result was that a final superheat of only from 8° to 10° was obtained, but at any rate the steam was dry instead of being wet as before, and an economy of 5 % was recorded over a year's working, after only two out of the seven boilers had been fitted with the superheaters.

The territory of the Company was extended in 1895 to include adjacent parishes to the north, and in 1896 a second generating station was constructed in the Commercial Road, Lambeth, on the south side of the Thames about midway between the Water-loo and Blackfriars Bridges, to supplement the output from the Maiden Lane Station. The new station started with six Belliss engines, each direct coupled to a 200 K.W. dynamo generating continuous current at 1,000 v. Steam was furnished by Hornsby water-tube boilers with 4,500 sq. ft. heating surface and an evaporative capacity of 12,000 lb. per hour. The 1,000 v. current was taken across Waterloo and Hungerford Bridges to sub-stations on the north side of the river, where its pressure was lowered to about 220 v. by 100 K.W. motor generators, to feed the low-tension distribution network. The high-tension mains had insulation of vulcanized bitumen, and were laid in cast-iron troughs filled with bitumen. The generating capacity of the Commercial Road Station eventually amounted to 4,400 K.W., made up of twelve units of 200 K.W. each, and two 1,000 K.W. units.

The Lambeth Station was shut down in 1909, its load being taken over by the new Bow Station, and the independent source of supply which it had constituted for theatres and public buildings, in accordance with the requirements of the London County Council, was then provided by Diesel engines installed in the sub-stations at Short's Gardens and St Martin's Lane.

In 1899 the Charing Cross Company obtained statutory powers to supply electricity in the City area, and although con-tinuous current was required by the consumers served by the Company, as it was in the Strand area, the only practical way of meeting the demand was to transmit alternating current at high voltage, from a power station at a considerable distance, to centrally situated substations where it could be converted and fed into the distribution system. A site of eight acres was obtained at Bow, some four miles from the City as the crow flies, and here, in 1900, the Company commenced the construction of a new power station, whence cables carrying three-phase current at 50 cycles and 11,000 v. were laid to six substations ranging from Fenchurch Street in the City to St Martin's Lane in the West End. The scheme attracted much attention at the time, for it was the first example in Great Britain of three-phase generation, transmission and conversion at 11,000 v.

The Bow Power Station, which was built and equipped according to the designs of Mr W. H. Patchell, the chief engineer of the Charing Cross Company, comprised an engine-room and a boiler-room, lying side by side, and each 75 ft. wide by about 300 ft. long. The station went into service in May 1902, the first equipment consisting of two 800 K.W. triple-expansion Belliss engines coupled directly to 11,000 v. alternators running at 230 R.P.M., and two 1,600 K.W. horizontal cross-compound Sulzer engines running at 83·3 R.P.M. with flywheel alternators on the centre of their crankshafts. Before these latter sets could be put into commission, the increase of the load made it necessary to order two more similar units, and in 1904 and 1905 the equipment of the engine-room was completed by a pair of 4,000 K.W. vertical three-crank compound Sulzer engines driving alternators from the ends of their crankshafts at 83·3 R.P.M. The 4,000 K.W. engines each had one H.P. cylinder $50\frac{1}{4}$ in. in diameter between two L.P. cylinders $70\frac{7}{8}$ in. in diameter, all with a stroke of $51\frac{1}{4}$ in. The engines stood 33 ft. 4 in. high from the floor level, and weighed 450 tons each without the alternator. The latter machines had 72 revolving poles, the fields being 25 ft. over the pole-tips. These immense units practically marked the culminating point of slow speed reciprocating machinery in power stations. Its doom was indeed already sealed, for in February 1905 there was a Parsons turbo-alternator with an economical rating of 4,000 K.W. at 1,200 R.P.M. and capable of an output of nearly 6,000 K.W.

already running in the Carville Power Station, and consuming only 15·4 lb. of steam per K.W.H., a figure that it is unlikely that the engines could approach.

The whole of the alternators were supplied by the Lahmeyer Company who also provided a pair of 300 K.w. dynamos, driven by triple-expansion Belliss engines, for the excitation of the main units. These were supplemented by a pair of 350 K.w. motor-generators and a storage battery. Excluding the exciters, the engine-room at Bow contained 16,000 K.w. of reciprocating machinery on a floor space of 22,500 sq. ft. All the engines exhausted into jet condensers, but, owing to the limited supply of circulating water available from the little creek at the side of the station, it was necessary to erect 16 circular steel cooling towers, 30 ft. diameter at the base and 85 ft. high, alongside the engine-room.

The boiler-house at Bow was as noteworthy as the engine-room for the size of some of its units. The boilers, which were all of Messrs Hornsby's manufacture, supplied slightly superheated steam at a pressure of 160 lb. per sq. in. They were set in pairs along the two sides of the house, with a steel chimney between each setting to deal with the gases from the adjacent boilers. The settings were steel-cased, which was a novelty at the time. The grates were all hand-fired, the coal being delivered from overhead bunkers. At one end of the house were a pair of small boilers, each with a heating surface of 4,590 sq. ft. and a rated evaporation of 12,000 lb. per hour. The design had a general resemblance to the land type Babcock boiler, with the difference that the tubes terminated in flat waterboxes, instead of in staggered headers. After this pair came five pairs of larger boilers of the same type, each with a heating surface of 8,100 sq. ft. and a rated evaporation of 24,000 lb. per hour. These were specially designed for the Bow Station, and were interesting because each was furnished with two grates which could be fired independently. The boilers were normally fired from the front only, but when there was a call for extra steam, as in case of sudden fog in the City, the second grates could be fired also, this being done from the gangways between the settings. In 1904 the first of the really large boilers

were installed. They were of the "Upright" type newly introduced by Messrs Hornsby, with groups of almost vertical tubes
expanded, above and below, into steam and water boxes. Two of
these boilers were combined to form a single unit, the largest in
the world at the time, with a heating surface of 21,700 sq. ft., and
grates aggregating 336 sq. ft., which could be fired from the
front and from each side. The rated capacity of each unit was
66,000 lb. of steam per hour, corresponding to an evaporation of
only 3·04 lb. per sq. ft., but the boilers were capable of steaming
at the rate of 100,000 lb. per hour when forced. Altogether six of
these large units were erected.

The Bow Station continued to run with its reciprocating
machinery until after the War. In 1919 a start was made to
modernize it. The reciprocating engines were removed one by
one to make room for turbines and the boilers were replaced by
others of Babcock and Wilcox manufacture, supplying steam at
270 lb. per sq. in. superheated to 650° F. Turbine machinery
rated at 74,250 k.w. in six units occupied less space in the engine-
room than the 16,000 k.w. of reciprocating plant, and the change
effected a reduction in the fuel consumption of the station from
4·7 to 1·68 lb. per unit generated.

THE BROMPTON AND KENSINGTON
ELECTRICITY SUPPLY CO., LTD.

On 24 January 1889 the House-to-House Electric Light Supply
Co., Ltd., which changed its name in the following August to that
of the Brompton and Kensington Electricity Supply Co., Ltd.,
inaugurated a service of electricity in the Brompton district from
a power station situated in the Richmond Road, Brompton. The
station was designed and erected by Robert Hammond, and gave
a supply on the single-phase alternating-current system at 83
cycles. Its initial equipment consisted of three 200 H.P. horizontal cross-compound non-condensing engines by John Fowler
and Co., Ltd., of Leeds, taking steam at 150 lb. pressure
from three Babcock and Wilcox boilers. The engines ran at
88 R.P.M. and each drove a 100 K.W. 2,000 V. Lowrie-Parker
alternator at 380 R.P.M. by seven 1½ in. cotton ropes from its

14 ft. flywheel. Each alternator drove its 3 K.W. series-wound exciter at 800 R.P.M. also by means of ropes. The voltage of the alternators was controlled automatically by a Lowrie-Hall regulator, which consisted essentially of a length of fine wire strung between two fixed abutments and kept taut by a weight suspended from its centre. The wire was connected across one of the stator coils, and acted somewhat in the manner of a hot-wire voltmeter, the variation of its sag being used to operate a lever which regulated the excitation current. It was claimed that by this means the variation of voltage could be kept within 1 %, but in view of the fact that the engines were controlled by hand manipulation of the stop-valve such close regulation may perhaps be doubted.

The principal objection to rope-driven generating sets of the type in question was the amount of room they took up. That they could render good service at times is proved by an affidavit sworn to by Mr C. J. Hall, the manager of the Richmond Road Station, to the effect that one of the sets there was started at 3.45 p.m. on 28 November 1890 and ran until 9.30 a.m. on 26 December, a total of 665·75 hours, or nearly 28 days without stopping. During this time the engine made 3,595,050 revolutions, and the alternator 15,938,055 revolutions.

The switchboard at Richmond Road was not designed for parallel running, but any circuit could be connected to any alternator by the insertion of a connecting block provided with four brass plugs. However, before the end of 1889, the practice of running the alternators in parallel every evening had been introduced by Mr Albert Gay, the first resident engineer of the station, so that the Richmond Road plant was one of the very earliest in the country to be operated in this way. The generator circuits and the feeder circuits were both protected by open-type fuses of copper wire, about 15 in. long. As the whole of the apparatus was working at 2,000 v., the switchboard, which was of the flat-back type with slate panels, was none too safe, and any mistake in switching was liable to cause unpleasant fireworks. The timber framing that carried the panels was, moreover, insufficiently rigid, and the whole board used to sway when a badly

PLATE XVIII. RICHMOND ROAD POWER STATION, 1889

From contemporary engraving in *The Engineer*

fitting plug switch was inserted or withdrawn. It remained in service, however, until 1900, when it was replaced by another of the same type, but of more robust construction, which in turn gave way in 1908 to a modern board with oil circuit breakers.

The current was taken from the station at 2,000 v. by rubber insulated cables drawn into iron pipes. At first every consumer had a transformer in his house for his own supply, but later on a low-tension network was established, fed by larger transformers at suitable points. The consumers' meters were of the electrolytic type, in which the quantity of current used was determined by the weight of copper deposited from a solution of copper sulphate. An electrolytic meter, by itself, would not of course work on an alternating-current circuit, as no deposition would take place. To get over this difficulty, the ingenious device of connecting a small 2 v. accumulator in series with the electrolytic cell was adopted. This caused a greater flow of current in one direction than in the other during every cycle of the alternations, and consequently there was a differential decomposition of the electrolyte proportional to the consumption of alternating current. The electrolytic cell was so proportioned as to allow an area of 4 sq. in. per amp. The same kind of consumers' meters were used on the alternating current supply of Eastbourne.

About 1895 the rope-driven generating sets were replaced by direct-coupled machinery, and in 1904 the station contained 12 Willans-E.C.C. and one Belliss-E.C.C. units, of a total capacity of 2,890 K.W., the boiler plant comprising two Lancashire boilers and 17 of the water-tube type. Most of the latter were by Babcock and Wilcox, but Stirling and Niclausse boilers were also in use. In 1911 the first turbo-generator was installed, this being a 500 K.W. reaction turbine of the Parsons type built by Messrs Willans and Robinson. The introduction of turbines involved the use of cooling towers, which were erected in steel tanks on the roof of the station. Another innovation was the adoption, in 1921, of oil-firing for the boilers, this being done mainly on account of the difficulty of dealing with coal and ashes on the site. In 1924 the capacity of the station had been raised to 8,000 K.W. by the installation of four more turbines. The plant continued to run until 1928, when it was shut down and a bulk supply of electricity taken by the Company from the mains óf the London Power Company.

THE CITY OF LONDON ELECTRIC LIGHTING CO.

In 1881 the Anglo-American Brush Electric Light Corporation made a contract with the City authorities for lighting New Bridge Street, Ludgate Hill and St Paul's Churchyard by means of arc lamps. The current was generated by belt-driven Brush arc-lighting machines in the Company's works in the Belvedere Road, and carried across the Thames by cables laid in cast-iron pipes over Blackfriars Bridge. This lighting was continued for several years, but it was not until 1890 that operations were commenced on a statutory basis. In that year Provisional Orders were granted to the Brush Electrical Engineering Co., Ltd., in respect of an area in the West Central district of London, and to the Laing, Wharton and Downs Construction Syndicate for an area in the City. The two Companies transferred their interests in February 1891 to the City of London Electric Lighting (Pioneer) Co., which had been registered the same month with a capital of £50,000. In July 1891 the City of London Electric Lighting Co., Ltd., was registered with a capital of £800,000. It acquired the assets of the Pioneer Company, paying back the shareholders the amount of their original investment with a bonus of 50 %, and with the consent of the Board of Trade it took over the Provisional Orders concerned.

The City of London Electric Lighting Co. inherited two power stations established by the Pioneer Company, one at Wool Quay, on Lower Thames Street, and the other at Meredith's Wharf, Bankside, on the south side of the river exactly opposite to St Paul's Cathedral. The Bankside Station had commenced supply on 12 June 1891, with an equipment of two pairs of 25 K.W. Brush arc-lighters and two 100 K.W. Brush single-phase alternators generating at 2,000 v. and 100 cycles. All were driven by belting from Brush engines, and steam was supplied by two Babcock and Wilcox boilers. Another boiler was put down the next year, and the whole of the generating machinery duplicated, the station as founded by the Pioneer Company being then complete.

The Wool Quay Station went into service on 19 December 1891 with two 60 K.W. Robey engines driving their alternators by belts, and two 25 K.W. B.T.H. arc lighters. The following year two Robey engines driving 140 K.W. alternators and a Fowler engine driving a 100 K.W. alternator were started up, as well as four more B.T.H. arc lighters. In January 1893 a 70 K.W. Willans set was added, but it had become evident that the site was a most unfavourable one for a power station. No further extensions were therefore made, but, bit by bit, the plant was transferred to Bankside or sold, and in January 1898 the Wool Quay Station was finally shut down.

The present Bankside Station, as distinguished from the Pioneer plant, dates from 1892, when a fine new engine-room and boiler-house were erected independently of the older buildings. The engine-room, about 230 ft. long by 50 ft. wide, started work in 1893 with two 200 K.W. Brush rope-driven alternators, two 400 K.W. direct-coupled alternating-current sets of the same make, and two 350 K.W. Willans engines each driving a pair of alternators from the ends of its crankshaft. The second alternator was coupled up by a magnetic clutch as half-load was reached. In the boiler-house, which ran parallel to the engine-room and was nearly the same length, were nine Babcock and Wilcox boilers. In 1895 the engine-room was extended to a length of 424 ft., and a pumping station was put down for condensing purposes, all plant having previously run non-condensing. In the same year the boiler-house was extended to a length of 300 ft., with a single row of 22 Babcock and Wilcox boilers.

By 1901 the boiler-house had been doubled in width and contained 46 Babcock and Wilcox boilers arranged on either side of a central firing aisle, while in the main engine-room there were ten B.T.H. alternators of a total capacity of 3,900 K.W. directly coupled to three-crank Willans engines, eight Brush alternators of a total capacity of 3,600 K.W. driven directly by open-type two-cylinder compound Brush engines, and two Ferranti compound engines driving 1,500 K.W. alternators at 150 R.P.M. making an aggregate of 10,500 K.W. capacity in twenty units.

Up to 1900 the whole supply was generated at 2,000 v., single-phase. In that year the construction of a new engine-room and boiler-house was started, in order to give a supply of direct current to the many printing works in the Company's area. These works did the printing of all the principal London daily papers, and up to that time had had their own generating plant. A third engine-room was built in 1903 for direct-current machinery for the same purpose. The first direct-current generators to be installed were three Allis Chalmers engines and dynamos, each of 1,000 K.W. capacity. These were followed by four Musgrave engines driving Westinghouse dynamos of 2,000 K.W. capacity in the second engine-room, and two further sets of the same make and capacity in the third engine-room. Steam for the direct-current plant was provided by Dryback Marine boilers in the new boiler-house, this type of boiler being selected as the newspaper printers did not care "to risk their supply by depending on new-fangled boilers of the water-tube type".

These extensions brought the capacity of the station up to 25,500 K.W. by the year 1907, 15,000 K.W. of this total being direct-current machinery. It is interesting to note that, whilst great difficulty was experienced in the early days in providing sufficient steam for a load of 10,000 K.W. from the 46 Babcock and Wilcox boilers originally installed, improved operating methods enabled these same boilers later on to furnish comfortably all the steam required for the 25,000 K.W. of plant eventually served by them.

The alternators at Bankside and Wool Quay were run in parallel from the start, and the two stations also ran in parallel. In paralleling the Ferranti alternators at Bankside, it was necessary first to get the crankshafts in opposite phase. This was done by means of an electical indicator attached to the shafts, the speed of the incoming engine being adjusted until perfect opposition was obtained. Until this was the case, the alternators were not paralleled electrically, as otherwise the cyclic irregularity of the engines would have caused a heavy surging current between the machines, even if the latter had remained in step.

The Ferranti engines, put down in 1898 and 1899, had to have the ingenious valve gear of their inventor much simplified before they could be got to work satisfactorily. They also swayed to such an extent that it was said that a man had to have "sea-legs" to remain on their platforms, while the vibration they caused resulted in the disintegration and ultimate fall of the statue of St George and the Dragon which adorned the frontage of the station. They ran, nevertheless, until 1911, being the last of the reciprocating engines to be taken out of the main engine-room when turbines came into use. The six Musgrave engines had an average life of well over 20 years each, the last ceasing work in February 1927.

The first turbo-alternator at Bankside was a 2,500 K.W. set which started in December 1910, and another of the same size went into commission about six weeks later. Other turbines followed almost yearly, and by 1920 there were seven turbo-generators in the main engine-room with an aggregate capacity of 19,500 K.W., generation being still single-phase alternating current at 2,000 V., and direct current at 450 V. In 1911 the first of two interesting direct-current turbo-generators was installed. The turbines were of 2,500 K.W. capacity running at 1,500 R.P.M. and each drove a pair of 1,250 K.W. 450 V. dynamos arranged in tandem. Although the collection of the heavy currents involved, at a speed of 1,500 R.P.M., was considered an ambitious project, the sets ran for many years and gave exceedingly good commutation.

In 1919 it was decided to change the system of generation from 2,000 V. single-phase, to 11,000 V. three-phase. At the same time the steam conditions were to be raised from the 150 lb., with a very slight superheat, which had sufficed until then, to 250 lb. pressure superheated to 660° F. Owing to the difficulty of obtaining plant so soon after the War, it was not until May 1922 that generation could be started under the new conditions. The new turbine sets were located in the main engine-room, which had already seen so many changes of plant. To accommodate the new boilers another boiler-house was built on the other side of the engine-room, and eventually the old boiler-house with its 46 small Babcock and Wilcox boilers, some of which had been working for a quarter of a century, was shut down.

New turbo-alternators and boilers were added annually, and by 1928 the old main engine-room, which in the early days had been uncomfortably filled with plant aggregating 10,500 K.W. capacity, contained plant of an aggregate output of 89,000 K.W., and Bankside ranged high in the annual list of station efficiencies issued by the Electricity Commissioners. By this time, with the exception of two stand-by steam sets, the whole of the steam-driven direct-current generators had disappeared, having been replaced by rotary converters run off the A.C. bus-bars. The second engine-room, built to contain the direct-current generators, was employed as a substation, while the third engine-room provided room for switch-gear and control apparatus.

Bankside became a selected station under the National Grid Scheme, and since 1934 has been operated to the requirements of the Central Electricity Board. Looking into the future, it is anticipated that the next development will be the erection of a super-station at Bankside. Land for this purpose has already been acquired on the east and west of the original site.

There are many stories of happenings at Bankside, which had an exceptionally difficult load to carry owing to the sudden fogs that were liable to descend on the City in the winter, but one of the most extraordinary relates to an incident that happened there some 20 years ago. The story would indeed be incredible were it not vouched for by a "cloud of witnesses", including Mr E. Harlow, the engineer at Bankside, who was present at the time. A small boy, 12 years of age, was playing about near the suction sump of the circulating pumps when he fell in. The accident was fortunately seen by a workman who gave the alarm, and the pump was instantly stopped. Meanwhile the boy had disappeared into the 36 in. suction pipe. This terminated in a sealed screen chamber, 150 ft. distant, the pipe making a double bend on the way. The screen chamber worked under a vacuum of 20 in. of mercury. An aircock was opened to release the vacuum, and the cover lifted by means of a screw and hand-wheel with the object of recovering the body of the child. It was 20 minutes after the accident before the cover could be partially raised, when the boy was heard to call out loudly, and he firmly grasped the hand that was extended into the chamber. He was found to be clinging to the screen with his head well out of the water, and was apparently none the worse for his experience. Nevertheless, in order to run no risks, he was wrapped in dry blankets and sent to the hospital, whence he was returned very shortly, as nothing could be found wrong with him. From the speed of the water flowing through the pipe, it is certain that the lad must have been totally submerged inside the pipe for a minute and a quarter, and after that he had spent about 20 minutes under a vacuum of 20 in. in the screen chamber. It is probable that his lungs, being full of air under normal pressure, were to some extent protected against the entry of water under the reduced pressure, but whatever may be the explanation of his escape, the facts are sufficiently remarkable.

THE COUNTY OF LONDON ELECTRIC LIGHTING COMPANY

The County of London Electric Lighting Co. was registered on 30 June 1891 with a capital of £100,000, one of its objects being "to produce and supply light, heat, sound and power by electricity, galvanism, magnetism or otherwise". Exactly what the promoters had in mind when they included the production of sound in their aims is an interesting subject for speculation, but they could have no doubt achieved it "by galvanism, magnetism or otherwise". Moreover, the relegation of the production of power to the last place on the list is significant of the views held at the time with regard to the uses of electricity.

In 1894 the Company changed its name to the County of London and Brush Provincial Electric Lighting Co., Ltd., and acquired sites for two power stations. One of these was on the City Road Basin of the Regent's Canal, whence a supply could be given to the St Luke's and Clerkenwell district, and the other was on the bank of the Thames at Wandsworth for the supply of the

Company's area on the south side of the river. Both stations commenced operations late in 1896. They were laid out for the generation of single-phase current at 2,000 v. and 100 cycles. The City Road Station started with twelve Babcock and Wilcox boilers and the same number of main generating sets. The latter consisted of Mordey and E.C.C. alternators of 150 K.W. capacity driven directly by Raworth "Universal" engines of 250 I.H.P. The engines were of the vertical enclosed compound type with superposed cylinders 19 and 32 in. diameter by 14 in. stroke. Each engine had only a single crank, and its speed was controlled by a flywheel governor which varied the cut-off. The design was originated by the Brush Company, and engines of the same kind were installed in a number of power stations in which the Company was interested, but the design did not meet with permanent success.

The Wandsworth Station was equipped on the same lines as the station at City Road. It commenced with six Babcock and Wilcox boilers and six "Universal" engines, all driving 150 K.W. Mordey alternators. It was found at Wandsworth that the alternators could be run in parallel in spite of the fact that they were directly driven by single-crank engines, the success being attributed to the absence of iron in the armatures of the Mordey alternators.

CHAPTER VI

SOME EARLY MUNICIPAL
POWER STATIONS

Self-love forsook the path it first pursued,
And found the private, in the public good.

POPE

ALTHOUGH the object of the Electric Lighting Act of
1882 was to place Local Authorities in a favoured position
as regards the supply of electricity within their areas, and
the effect of it was to paralyse the development of Company
undertakings generally, several years had to elapse before any
Municipality took the risk of embarking on any public supply of
electricity. Very many of them obtained Provisional Orders
under the Act from the Board of Trade, but apparently the motive
of such action was more often the desire to keep Companies out of
the field than to take any further steps themselves towards the
establishment of an undertaking of their own. In due course a
certain number took over Company undertakings, especially in
some of the larger towns, but others founded businesses for
themselves, and occasionally showed a praiseworthy enterprise
in the practice they adopted. The Corporation of Portsmouth, for
example, installed a steam turbine in 1894, in advance of any of
the large Companies in the Metropolis; Wolverhampton led by
generating and transmitting continuous current at 2,000 v. in
1895; while the Local Authorities of Shoreditch and elsewhere
deserve credit for their attempts to generate electrical power
from steam produced by the incineration of ashbin refuse.

BRADFORD MUNICIPAL STATION

The first Municipal Authority to enter the electric lighting
business was the Corporation of Bradford in Yorkshire. They
were granted a Provisional Order in 1883, but did not take any
hasty action in connection with it, for it was not until 1887 that an
undertaking was planned for them by Mr J. N. Shoolbred, the

consulting engineer to the Corporation. A heated controversy at once arose concerning the merits of the scheme, the proper site for the power station, and so on. To get the matter settled the Corporation asked Dr John Hopkinson to visit the town, inspect the alternative sites, examine the plans and specifications, and give his advice on the whole question. He reported generally in favour of Mr Shoolbred's scheme, which was consequently proceeded with, and on 20 September 1889 the first Municipal electrical undertaking was ready to give a supply to consumers. The power station was situated in Bolton Road, and the whole of its electrical equipment was furnished by Messrs Siemens, Bros. and Co., who entered into an arrangement with the Corporation whereby they should manage the plant for a month or two after it was first put into commission.

The system of supply decided on at Bradford was two-wire direct current at low-tension, the pressure on the mains being 115 v. The initial equipment of the station consisted of three 100 K.W. Siemens shunt-wound dynamos, of which two were coupled directly to a pair of two-crank Willans engines of 150 H.P. each, and the other was driven by a Marshall engine of the same power. The latter engine was of the two-crank compound vertical type, with steam-jacketed cylinders 12 and 19·5 in. diameter respectively, and a stroke of 18 in. It was fitted with a Hartnell governor controlling an expansion slide valve in the H.P. steam chest, and with hand-controlled Meyer valve gear for the L.P. cylinder. Steam was supplied by three Lancashire boilers, each 7 ft. diameter by 28 ft. long, working at 120 lb. per sq. in. As usual with such boilers there was a battery of Green's economizers, and the feed water, before entering them, was heated by the exhaust steam from the engines in an open-type heater.

For the first few months the supply of electricity was available only from an hour before sunset until 11 p.m. The service was then extended so as to start at 10 a.m. every morning. In 1890 the machinery was increased by the addition of a 300 I.H.P. Willans-Siemens unit, the engine of which was of the three-crank type, the first of its kind to be made. This set was duplicated the following year, and two new Willans-Siemens sets,

PLATE XIX. BRADFORD POWER STATION, 1892

From contemporary engraving in *The Electrical Review*

each of 80 I.H.P. together with a storage battery of 70 Crompton-Howell cells having a capacity of 1,000 amp.-hours, were installed about the same time. The station then contained engines aggregating 1,200 H.P., and a full 24 hours' service was being given. Four more Lancashire boilers had been put in to supply the additional engines, and in 1892 a Babcock and Wilcox boiler with an evaporative capacity of 4,200 lb. per hour was added to the plant.

The load on the Bradford undertaking grew with great rapidity, and in 1895, in order to increase the capacity of the distribution system, Mr Sydney Baynes, who was the electrical engineer to the Corporation at the time, converted the network to the three-wire system with 230 v. between the outers. Balancing was effected by a battery across one side, and a generator across the other. The important consequence of the change to three-wire distribution was that Mr Baynes took advantage of it as a means of bringing about a general increase of the voltage in Bradford. All new consumers were connected across the outers, so that they were supplied at a pressure of 230 v. instead of 115 v. As it was not very easy to obtain satisfactory 230 v. lamps in those days, the action of the engineer gave rise to much discussion, but, whatevery may have been meant when the voltage of supply was originally defined, the regulations were so worded that there was nothing contrary to the letter of them in what was done at Bradford. Within a few years the pressure of the network was raised to 460 v. across the outers, and all domestic lighting had thenceforward to be done by 230 v. lamps.

In order to cope with the increasing demand for electricity, a new power station was built at Valley Road, its foundation stone being laid on 5 June 1896. This station went into operation the next year with an initial equipment of four three-crank Willans engines driving Siemens dynamos. Two of the sets had a capacity of 375 K.W. at 500 v. and 300 R.P.M., and the other two were of half this output. Steam at a pressure of 180 lb. was provided by two Marine boilers built by Messrs John Brown and Co., each with an evaporation of 12,000 lb. per hour. The boilers, which were 9 ft. 9 in. diameter by 10 ft. 6 in. long, were fitted with

Serve tubes, and supplied with preheated air on the Ellis and Eaves induced-draught system. The engines exhausted into a surface condenser with 2,500 sq. ft. of tube surface. The auxiliaries were electrically driven, and included two single-acting air-pumps which removed the condensate.

THE SAINT PANCRAS UNDERTAKING

The first Local Authority in the Metropolis to embark on the business of electricity supply was the Vestry of Saint Pancras. The Vestry had obtained a Provisional Order for the purpose in 1883, the object, however, being to keep Company undertakings out of the territory rather than to take any steps to provide a supply of electricity themselves. The matter was therefore allowed to rest until 1890, by which time it had become evident that the policy of "masterly inactivity" could no longer be maintained in view of the intentions of the Board of Trade that a supply of electricity should be available to all who needed it. The Vestry consequently decided to take action and instructed Professor Henry Robinson, as their consulting engineer, to prepare plans for an undertaking. In order to make sure that they were getting the best advice, the Vestry also obtained professional opinions from Dr John Hopkinson, then President of the Institution of Electrical Engineers, and Sir William Preece, before coming to a decision as to the system to be adopted. This was the low-tension direct-current three-wire system with batteries. A site for the power station was acquired in Stanhope Street, near Regent's Park, and the foundation stone laid on 5 November 1890.

In their efforts to attract customers the Vestry showed commendable business enterprise. They organized an Electrical Exhibition in the Borough, thus awakening the interest of potential consumers in electrical apparatus, and at the same time benefiting by the arguments of the exhibitors, whose advantage lay in popularizing the use of electricity. The Vestry also adopted a low scale of charges for current, with the consequence that within six months of the commencement of supply on 9 November 1891, the whole available output of the station was being disposed of.

PLATE XX. ST PANCRAS POWER STATION, 1891

From contemporary engraving in *The Electrical Review*

The St Pancras Power Station was considered a model of its kind at the time. It was designed to supply current for 10,000 incandescent lamps of 16 c.p. and ninety 10 amp. arc lamps for street lighting. The engine-room, which was 106 ft. long by 26 ft. wide, contained nine main generating sets, each consisting of a six-pole Kapp dynamo with an output of 680 amp. at 112–130 v., driven directly by a two-crank triple-expansion Willans engine. There were, in addition, a pair of dynamos giving 90 amp. at 540–575 v., also driven by Willans engines for arc-lighting, though they could also be used for charging the batteries in series. Each was capable of supplying current to nine arc circuits, with ten lamps in series per circuit.

Steam was furnished at 170 lb. pressure by five single-drum Babcock and Wilcox boilers with a heating surface of 1,619 sq. ft. and a rated evaporation of 5,000 lb. per hour. The boilers were served by a 90 ft. brick chimney, 5 ft. square internally. A pipe extending for 50 ft. up the side of the chimney carried off the exhaust steam when the engines were exhausting to atmosphere. The normal exhaust of all engines was into jet condensers drawing water from an underground tank of 170,000 gallons capacity. The hot water was returned to the top of the tank, and pumped thence to a cooling arrangement on the top of the building, 50 ft. up, whence it flowed back to the bottom of the tank. The cooler consisted of a zigzag construction of corrugated metal sheets, about 14 ft. high, over which the water flowed at the rate of 10,000 gallons per hour. The clouds of steam that rose from the cooler day and night formed quite a characteristic feature of the station.

Four batteries, each of 60 cells with a discharge rate of 60–75 amp., were maintained in the station to equalize the voltage, and another pair of batteries was placed in an underground sub-station three-quarters of a mile away. The distribution of the current to the consumers was carried out by the three-wire system with 240 v. across the outers. The mains were built up of bare copper strip 1·5 by 0·125 in. in cross-section, carried on insulating bars in conduits. The consumers were provided with Aron or Ferranti meters, and were charged at the rate of 6d. per

K.W.H. After the station had been running a few months, a tariff of 3d. per K.W.H. was introduced for daylight supplies, this current being registered on a separate meter.

In September 1893 the Vestry authorized the completion of the Stanhope Street Station by the installation of three more units of 90 K.W. with another three boilers, and they decided at the same time to proceed with the construction of a second power station in King's Road. The interest of the new station lay in the fact that it was designed to operate in conjunction with a refuse destructor. It was put into service towards the end of 1895. The destructor plant was laid out for 18 cells, but only six were built to commence with. Each cell was 5 ft. 3 in. long by 5 ft. wide and 2 ft. 6 in. high. The hot gases from the destructor passed into a flue running the whole length of the boiler-house behind the boilers. There were three of these, of the Lancashire type 7 ft. diameter by 30 ft. long working at 125 lb. pressure, and with a rated evaporation of 5,500–7,000 lb. per hour. The boilers could be fired with coal in the ordinary way, but when steam was being raised by the destructor gases, these were diverted from the main flue mentioned above, and passed to the stack by two circuits in parallel. One path was backwards through the internal flues of the boiler to a cast-iron downtake at the front, and so to the stack, while the other passed underneath the boiler to the stack. The downtake was shut off when the boiler was fired with coal, and the flue gases then took their normal path. A 240 tube Green's economizer, with a surface of 2,500 sq. ft., was installed for heating the feed water, which was previously treated by a Doulton softener and measured by a Kennedy meter. The economizer was fitted with a steam drum, gauge glasses, low-water alarm, and in fact could operate exactly like a boiler; for a return pipe connected the water space of the drum to the inlet of the economizer, so that a circulation through the tubes could be maintained. The steam produced escaped through a reducing valve to the ring main of the station.

The generating plant at the King's Road Station consisted of three two-crank compound Belliss engines developing 200 B.H.P. at 350 R.P.M., each mounted on a bed-plate between two 65 K.W.

120 v. two-pole Crompton dynamos coupled to the two ends of the crankshaft. The engines exhausted either into three Ledward ejector condensers, or to atmosphere through a 12 in. vertical pipe 50 ft. high. The cooling water arrangements were very similar to those at Stanhope Street, with a 120,000 gallon underground tank, and a cooler of corrugated sheets with a surface of 1,722 sq. ft. on the station roof.

When the King's Road Station was designed, it was intended to distribute the current from it on the five-wire system, as in Manchester, and a considerable length of five-wire mains was laid. When Mr Baynes took charge of the undertaking in 1895 however, he decided to give a three-wire supply and to raise the consumer's pressure to 230 v., as he had recently done at Bradford, and developments therefore proceeded along these lines.

PORTSMOUTH MUNICIPAL STATION

To the Portsmouth Municipality belongs the credit of being the first Public Authority to employ a steam turbine in its power station. Unlike the Company stations at Newcastle-on-Tyne, Cambridge and Scarborough, which were designed with nothing but turbine machinery, the Portsmouth plant started with both reciprocating and turbine units, and by so doing was able to provide the first demonstration of the ease with which slow-speed engine-driven alternators could be operated in parallel with turbine-driven machines. The Portsmouth Station was officially opened on 6 June 1894, and then contained a Parsons turbine with a capacity of 150 K.W. at 3,000 R.P.M. and a pair of slow-speed reciprocating sets, each rated at 212 K.W. at 96 R.P.M. All the machines generated single-phase current at 2,000 v. and 50 cycles. The slow-speed flywheel alternators were built by Ferranti, and were notable as being his first alternators to be designed with a revolving field, his earlier machines all having stationary fields and revolving armatures. They were driven each by a horizontal side-by-side compound condensing engine with Corliss valves, manufactured by Yates and Thom of Blackburn. The engines, which had cylinders of 15 and 28 in. diameter by 36 in. stroke, were in all essential respects similar to the type in general use at the time in the cotton mills of Lancashire, where they had gained a high reputation for economy and reliability. They were fitted with jet condensers, housed in the foundations, and served by air pumps 19 in. diameter by $11\frac{1}{4}$ in. stroke, driven by bell-crank levers from the L.P. cross-heads. Salt water from the harbour was used for condensing purposes, the whole of the heat in the exhaust steam being lost in the sea, as town water was employed for feeding the boilers. The cold feed water was heated by live steam before it entered the economizer. The boiler-house contained four Lancashire boilers working at 160 lb. pressure. They were fitted with Vicars' mechanical stokers, and worked in conjunction with a Green's Economizer in the main flue. Every boiler and every generating set had a duplicate connection with the steam range.

The Parsons turbine was of the radial flow type, similar to the Cambridge and Scarborough machines, but was larger than they were, being, in fact, the most powerful radial-flow turbine constructed by Parsons for duty on land, as the recovery of the rights under his patents enabled him to revert the same year to his original parallel-flow design.

The switch-gear at Portsmouth was of the simplest possible description. One pole of every alternator was earthed permanently, switching being done entirely on the other pole. A similar arrangement was adopted for the feeder circuits, the outer conductor of each concentric cable constituting a feeder being permanently earthed, and a switch being placed in the inner conductor only. There was thus only one switch per generator and one per feeder. This enabled the whole of the switch-gear to be accommodated on a board 6 ft. wide by 3 ft. 6 in. high. The switches were of the single-blade knife type, and provision was made for synchronizing by arranging that the blade of every generator switch should touch a "synchronizing contact" when brought into a half-closed position. As soon as synchronism had been obtained, as shown by the usual transformer and lamp, the switch was pushed home, leaving the synchronizing contact and entering the main contact.

PLATE XXI. PORTSMOUTH POWER STATION, 1894. SHOWING TURBO-GENERATOR IN FOREGROUND

Rectified current was used for arc lighting, since this combined the advantages of alternating current for the lamp mechanism with the superiority of unidirectional current for the crater formation, and therefore for the distribution of the light. Current for arc lighting was taken from the switchboard to the fixed coil of the constant-current transformer of a Ferranti rectifier. This was a large flat coil sandwiched in between two moving coils of similar shape which formed the secondary winding. The moving coils were suspended by a system of rods and balanced levers in such a way that they both approached or receded from the fixed coil, equally and simultaneously. When there was no current passing, the two moving coils both touched the fixed central coil. As the current in the latter coil increased, it exercised a greater and greater repulsion on the moving coils, with the consequence that they moved away from it. The device was essentially a transformer so arranged that its magnetic leakage was proportional to the load, and its effect was to maintain a substantially constant current at a variable voltage in the secondary circuit. The current was, of course, alternating. It was rectified by two commutators driven directly by a small synchronous motor mounted on top of the framework of the machine. The apparatus did its work satisfactorily, but it used to be humorously said that it gave as much light as one of the arc lamps. The rectified current was taken to another switchboard, where it was distributed amongst the various arc lighting circuits, which carried some forty or fifty arc lamps in series.

For incandescent lighting, the station pressure of 2,000 v. was reduced to 100 v. by Ferranti transformers housed in street boxes. To lessen the transformer losses at times of light load, a man used to go round the street boxes armed with a key, by means of which he could connect the two primary coils of a transformer in series and simultaneously could put the two coils of the secondary winding also in series. When the load was high, both sets of coils were connected in parallel by a reversal of the operation.

Dr Gerald Stoney, who was in charge of the erection of the Parsons turbine at Portsmouth, relates a story told him by

Ferranti of an amusing incident in connection with the official opening of the station. In the evening there was a municipal banquet, and afterwards a number of civic dignitaries and their guests went down to inspect the station. When they arrived the engine-room was brilliantly lighted by arc lamps supplied with current from the Parsons unit. One of the slow-speed engine-driven sets was then started, and its massive flywheel, crowned with its conspicuous row of field coils, began slowly to gather speed. As the set approached full speed, the visitors were astonished to see the rim revolving slowly backwards, although the piston rods and cranks kept up their normal pace and direction. While they gazed, wondering, the rim stopped and then proceeded to move leisurely forward. It then stopped again, and commenced a backward motion, finally stopping altogether! Some of the guests could not conceal their amazement, but others, more cautious and remembering the banquet they had enjoyed, just gripped the hand-rails and said nothing. The explanation of the phenomenon is, of course, simple. The light produced by the rectified current supplied to the arc lamps, although apparently continuous, consisted in reality of 100 flashes per second, each flash providing a momentary glimpse of the flywheel. When the latter was revolving synchronously with the turbine speed, successive magnet coils would occupy exactly the same position in space at every glimpse, and the flywheel rim would consequently appear to be stationary. If the engine then gained slightly in speed on the turbine, each coil would be slightly ahead of the position occupied by its predecessor during the preceding glimpse, and the rim would then appear to be slowly advancing.

THE HAMPSTEAD UNDERTAKING

A public supply of electricity in Hampstead was first given by the London and Hampstead Battery Co., Ltd., which commenced operations on a very small scale in 1892. The same year the Hampstead Vestry obtained a Provisional Order, and in 1893 after much discussion it was decided to follow the example of St Pancras and to establish a Municipal Station. This was formally opened on 1 October 1894, thus antedating the Municipal supply of Ealing by a few days. The power station was erected in the Vestry's stoneyard, off the Finchley Road. The engine-room, which measured 58 by 42 ft., was on ground level, but the boiler-house, of practically the same dimensions, had its floor 9 ft. below the ground level, the object being to enable coal to be brought direct to the boiler fronts from a storage vault underneath the roadway. The practice of placing boilers in an excavation in this manner so as to facilitate the coal supply, was common in the early days of the industry, in spite of the obvious disadvantages it involved on the score of drainage, and the greater difficulty of disposing of the ashes.

The steam-raising plant consisted of four Lancashire boilers 28 ft. long by 8 ft. diameter, with a rated evaporation of 5,000 lb. per hour and a working pressure of 150 lb. per sq. in. They were fitted with mechanical stokers, to the hoppers of which the coal was brought by a conveyer. The feed water, drawn from the mains or from a 10,000 gallon storage tank, was heated in an exhaust steam feed heater and delivered to the boilers through a Green's economizer of 192 tubes, by means of a pair of Worthington pumps. The flue gases were discharged through a brick chimney 130 ft. high and 8 ft. in diameter internally.

The engine-room contained four main generating sets, all consisting of Willans engines directly coupled to Siemens single-phase alternators. Current was produced at 2,000 v. and 90 cycles. All the engines worked non-condensing, some of their exhaust steam being used in the feed-water heater and the rest discharged to atmosphere through a 50 ft. pipe. The largest unit had an alternator of 200 k.w. capacity driven by a three-crank engine; two more units of 100 k.w. each and one of 20 k.w. completed the equipment of alternating-current plant. The alternators were operated in parallel, and their exciting current was provided by an 18 k.w. steam-driven dynamo which also supplied the station lighting. The public lighting was carried out by 22 Siemens arc lamps, of 2,000 c.p. each. They were connected in series, and for their service two 13 k.w. high-voltage direct-current sets were installed. The ordinary alternating-current supply was transmitted to seven transformer pits constructed under the pavements. Each pit contained a pair of 25 K.W. transformers to reduce the feeder pressure of 2,000 v. to 210/105 v. for the three-wire distribution network.

THE ISLINGTON VESTRY UNDERTAKING

Among the earlier Municipal Authorities to provide for the electrical needs of the inhabitants of their area was the Vestry of Islington, which laid the foundation stone of a power house of its own on 27 December 1894. The station, which was situated in Eden Grove, Holloway Road, was formally opened by the Lord Mayor and Sheriffs of London on 4 March 1896, although it had been supplying current since the end of January. The boiler-house, 109 ft. long by 56 ft. wide, contained six boilers, working at 150 lb. pressure and all with a rated evaporation of 5,000 lb. per hour. Four of the boilers were of the Lancashire type, 28 ft. long by 7 ft. 6 in. diameter. The others were Babcock and Wilcox boilers with a heating surface of 1,600 sq. ft. and a grate area of 30 sq. ft. They were fitted with soot blowers which were something of a novelty at the time. All the boilers worked normally with natural draft, although Meldrum blowers were provided in case of need. The flue gases were discharged by a chimney 180 ft. high, with an internal diameter decreasing from 14 to 12 ft.

Coal was brought to the station by rail, the 10-ton trucks being passed over an Avery weighbridge and then emptied by end-tipping. The Eden Grove Station had the distinction of being the first in Great Britain to handle coal in this way.

The engine-room was 80 ft. long by 88 ft. wide. It was equipped with four main generating units, each consisting of a horizontal compound slow-speed Adamson non-condensing engine of 250 I.H.P. driving a single-phase alternator of 125 K.W. capacity generating current at 2,000 v. and 50 cycles. Each engine had cylinders 15 and 24 in. diameter by 36 in. stroke, and ran at 95 R.P.M. Although the engines were all similar, the same indecision was shown as regards the alternator as had been manifested in the selection of the boilers, and with less excuse. Two of the alternators were driven at 375 R.P.M. by ten 1·25 in. ropes passing over the 14 ft. flywheels of their respective engines. The other two alternators were of the slow-speed flywheel type, mounted directly on the crankshafts of the engines and running therefore at 95 R.P.M. One of them was built by Ferranti, and was of the same design as the machine he had supplied to Portsmouth in 1894, while the other slow-speed alternator, as well as the two rope-driven machines, were manufactured by Messrs John Fowler and Co. of Leeds. Each alternator had its own 50 v. exciter driven at 450 R.P.M. by ropes from the alternator shaft. No provision was made for interchanging the exciters or for running them in parallel. A small engine-driven dynamo of 10 K.W. supplied current for the purposes of the station.

The four main alternators worked in parallel on one set of bus-bars, and four 2,000 v. feeders went out from another set of bus-bars. Between the two sets of bars was a Kelvin wattmeter balance, to measure the total output of the station. The voltage of the feeders was reduced to that of the network by transformers housed in street boxes.

THE SHOREDITCH DESTRUCTOR STATION

One of the primary duties of a municipality is to dispose of the household refuse of its citizens in a hygienic way. The practice of burning it in a properly designed refuse destructor was introduced some 60 years ago, and the heat of combustion was used to raise steam for the production of draught and other requirements of the plant. It was apparent that under favourable conditions, much more steam could be produced than was needed about the destructor, and there was something very attractive in the idea of combining a Municipal Destructor with a Municipal Power Station, so as to make use of the steam for the generation of electricity. Priority in this matter has been claimed for the Oldham Corporation, which took the excess steam from its destructor to its adjoining power station in March 1896, but as early as 3 October 1894 the Ealing Local Board had put the principle into practice. On that day the Board formally opened a small power station designed to serve 70 arc lamps and the equivalent of 8,000 incandescent lamps of 8 c.p. The interest of the station lay not only in the fact that it was arranged to make use of waste heat from a destructor plant, but also because sewage was to be employed as cooling water for the condensers. The waste heat was estimated to raise sufficient steam to develop 50 H.P. The station was equipped with three Lancashire boilers fitted with McDougall stokers for burning coal to supplement the heat from the destructor. In the engine-room were three Browett and Lindley high-speed compound engines direct-coupled to Siemens arc-lighting dynamos developing 10 amp. at 1,800 v. and running at 300 R.P.M. The incandescent lighting was provided for by two pairs of 2,000 v. 40 cycle alternators, one pair being 60 K.W. machines, and the other pair 30 K.W. machines.

A much more important application of the heat from destructors to electric lighting was made later at Shoreditch. The local Vestry had to pay contractors nearly £2,000 annually for the carting away of some 20,000 tons of ashbin refuse. The possibility of burning this in a destructor and generating electricity cheaply from the steam it produced, seemed to afford an excellent solution to the problem of refuse disposal, so, in 1896, the Vestry built a combined power station and destructor plant in Coronet Street, which was officially inaugurated by Lord Kelvin on 28 June 1897. The destructor house was 80 ft. long, and was served by a 150 ft. chimney. It contained a row of six Babcock and Wilcox boilers working at 160 lb. per sq. in., sandwiched between two rows of destructor cells, six on each side of the boilers. Each boiler was heated by two cells, and any boiler could be cut out by means of dampers. From 8 to 12 tons of refuse could be burnt in each cell per day, draught being provided by fans and steam jets. The supply of steam was equalized by a Druitt-Halpin storage drum, 8 ft. in diameter and 35 ft. long, in which the feed water was raised to the full steam temperature before entering the boiler.

In the engine-room were three three-crank Willans engines direct coupled to E.C.C. dynamos with a rated capacity of 165 K.W. at 350 R.P.M. There were also three smaller units of the same type, with an output of 70 K.W. each at 470 R.P.M. The larger machines generated continuous current at 1,100 v., which was transmitted to substations for conversion to a lower pressure, according to the "Oxford System". There was a two-wire distribution network at 165 v., which was fed directly by the smaller sets at the generating station. The high-voltage current was sent to three substations, the first of which, situated in Great Eastern Street, was equipped with three motor generators taking 1,100 v. current and each giving an output of 400 amp. at 165 v. for the network. At the power station there was also a motor generator having an output of 60 K.W. at either voltage to give flexibility to the system. Some of the high-voltage current was used for arc lighting without any transformation, nineteen 12·5 amp. arc lamps being run in series on a 1,100 v. circuit.

The Shoreditch Station gave rise to a considerable controversy concerning the real value of refuse as fuel. The opponents of the system contended that so much coal had to be used as well that the value of the refuse was practically negligible, and that it would be sounder engineering to keep destructor plants and electric power stations quite separate and independent of one another. Although it has to be admitted that a destructor plant is not a very nice neighbour for a power station, and that its ability to contribute steam must be discounted by the smallness and the irregularity of the supply, there was, all the same, a good deal that could be said for the combination when power stations were still on a very small scale. In connection with the Shoreditch controversy Mr H. E. Kershaw, the Chairman of the Shoreditch Lighting Committee, sent a letter to the *Electrical Review* in November 1897 in which he stated: "We are absolutely raising from our ashbin refuse sufficient steam to drive our electrical plant, giving a maximum output at our heavy load, of 250 K.W., and this we are raising solely from ashbin refuse." This appears categorical enough, and to it may be added the statement that during the year ending March 1899, a total of 1,031,348 K.W.H. was supplied by the Shoreditch Power Station, of which about 70 % was generated from refuse, a load of 400 K.W. being often carried on refuse only. Such figures are, of course, quite insignificant to-day, as the demand for electricity has increased ten thousandfold, and the heat from all the refuse that a community could make would have no appreciable effect upon the fuel consumption of a modern power station.

CHAPTER VII

THE BATTLE OF THE SYSTEMS

Where men of judgment creep and feel their way
The positive pronounce without dismay.   COWPER

WHEN central stations for the public supply of electricity began to succeed the arc-lighting installations of the early eighties of the last century, generators for producing alternating and direct current were both available. Both kinds of current were used for arc lamps and each had its advocates. This difference of opinion as to the merits of the rival forms of current was carried into central station work, where it soon became acute owing to the far greater importance of the question in the new field that was being opened up. Pioneers like Ferranti, Mordey and Gordon made no secret of their belief in the eventual triumph of the alternating-current system and devoted all their energies to its development, while on the other side the equally outstanding names of Crompton, Hopkinson and Kennedy were to be found amongst those who were convinced that the commercial future of electricity supply was bound up with the continuous-current system.

Now that the matter has long been settled in favour of alternating current, at any rate so far as public supplies are concerned, it is perhaps a little difficult to realize how strong a case could be made for continuous current during the period when the controversy was at its height. One has to remember that it was a commonly held opinion in those days that each small district would naturally be served best by its own power station, situated near the centre of the load, and confining its supply to customers within a radius of a mile or less. Ferranti had had greater vision when he established a great station at Deptford in 1889 for the supply of London, but the results of this venture, when compared with those attained by Crompton about the same time with his Kensington Court Station, were hardly calculated to impress either shareholders or consumers with the superiority of alter-

nating-current supply from a distant station over a local continuous-current system with batteries as a stand-by. The great
majority of the Companies participating in the supply of London
put their faith in direct-current systems, the most notable exception, apart from the London Electric Supply Corporation, being
the Metropolitan Electric Supply Co., but even this Company,
although relying almost exclusively upon alternating current,
had no fewer than five separate power stations for the service of
the area under its jurisdiction. This action affords evidence of the
parochial outlook which prevailed, even in the minds of engineers
responsible for the conduct of an ambitious programme of supply.
Some of them argued that a small station could be as efficient
and economical as a large one. Even as late as 1894, Professor
A. B. W. Kennedy, who was consulting engineer to the Westminster Electric Supply Co., stated at a meeting of the Institution
of Electrical Engineers, with reference to the sizes of power
stations: "However, when they came up, at a very rough ap-
proximation, to 3,000 H.P., or possibly 4,000 H.P. in the station,
they had got to a point beyond which practically they were not
going to reduce anything in their cost per unit." The view that
each parish should have its own power station, which underlay
the Electric Lighting Act of 1882, may no longer have been
openly defended, but the principle of independent local generating
stations placed at points where the demand was concentrated,
still had many adherents. Those who believed in this principle
could, of course, see no advantage in the facilities of high-voltage transmission afforded by the alternating-current system.
Continuous currents at a voltage suitable to the consumer could
be sent over the comparatively short distances involved without
undue loss, and this being the case, the most useful characteristic
of alternating currents was of no benefit. Alternating current
was no better for incandescent lighting than continuous current,
it was inferior to it for arc lighting, and it had the serious drawback that it could not be used for driving motors, for the first
alternating-current systems were all single-phase and no suitable
motors were yet available.

Another obstacle to the use of alternating current in the early days was the high cost of generation. An alternating-current station might consume two or three times as much coal per K.W.H. of output as a continuous-current station, operated in conjunction with batteries, although the distribution losses of the latter might be somewhat higher. The average coal consumption of alternating-current stations was stated by the *Electrical Review* in 1891 to have been from 20 to 25 lb. per K.W.H., as compared with about 10 lb. for direct-current plant with storage batteries. This was partly due to the difference in the efficiency of the machinery, for alternating-current engineers held on to rope and belt driving for some years after high-speed direct-coupled sets had become general for continuous-current work. But there is no doubt that it was mainly on account of the fact that the engines and boilers of an alternating-current plant had to be of sufficient capacity to carry the short daily peak load of the lighting demand, and were consequently running under very uneconomical conditions during the rest of the 24 hours. The machinery could never be shut down during periods of light load, as in a battery station, but had to be kept running, however small the demand. Conditions were made worse by the fact that for a good many years it was found practically impossible to operate alternators in parallel. It was therefore the custom to segregate the machines, each being used to supply only its own feeder or group of feeders, independently of the other units. This practice made it necessary to keep a number of machines running at times when the total load on the station was within the capacity of a single unit, or otherwise to interrupt the supply momentarily whilst transferring feeders from one machine to another. The argument against alternating-current supply which probably had the greatest weight, was the fact that the continuity of the service to consumers depended absolutely upon the uninterrupted operation of the machinery. Any failure of an engine or a generator in an alternating-current station was liable to cause a sudden and total extinction of the lamps in the consumers' houses, and such accidents were far from uncommon, as the generating units were not so reliable as they might have been, and in many instances troubles with belts or ropes had to be contended with.

From what has been said it will be seen that the advocates of the continuous-current system had a very strong case in the early days of the power station industry. By the use of batteries they could guarantee reliability of service, at any rate so far as concerned anything happening to their machinery. The station could, if necessary, be shut down for hours together, leaving the batteries to maintain the supply. Less stand-by plant was therefore required, as there was a daily opportunity of effecting repairs to the generating sets. Furthermore, whether these sets were engaged in charging batteries or supplying the load in parallel with them, they could always be operated at an economical rating, and consequently the efficiency of generation was high. Losses, of course, were incurred in the batteries, which probably failed to deliver to the mains more than about 80 % of the power put into them, while their capital and maintenance charges had also to be taken into consideration. Their rate of depreciation also was high, and it was even humorously contended by advocates of the rival system that one of the greatest merits of alternating current was that it precluded the employment of batteries! It has nevertheless to be admitted that the advantages of ensuring continuity of service, of providing a reserve supply of current for times of emergency, of enabling periods of light load to be dealt with economically, and of allowing the generating plant to be operated under the best conditions, were sufficiently important to place the direct-current battery system in a very favourable position. Indeed, had the business of electricity supply been destined to continue on the same scale as that of the earlier undertakings, there is little doubt that the direct-current system would have been generally adopted.

The direct-current engineers certainly did all they could to adapt their system to greater outputs and more extended areas. The 100 v. two-wire method of distribution was quickly superseded by the three-wire method, patented by Hopkinson in 1882, with 200 v. across the outers, and the pressure was raised to 400 v. as soon as 200 v. incandescent lamps were commercially available. At Manchester in 1893 the Hopkinson principle was carried to the extent of laying down a five-wire distribution network with 400 v. across the outers, a system which had already been in use for several years in Paris. Enterprise was also shown in other directions. Transmission pressures of 1,000–1,500 v. or more were adopted at Colchester (1884), at Chelsea (1889), at Oxford (1892), at Wolverhampton (1895) and at Lambeth (1896), the voltage being reduced for the consumers either by batteries, continuous-current transformers or motor generators. One of the most notable developments in connection with continuous-current working was the installation of plant operating on the constant-current Thury system by Mr J. S. Highfield at the Willesden Station of the Metropolitan Electric Supply Co., Ltd., in 1910. The generators, which were electrically connected in series, each produced a direct current of 100 amp. at 5,000 v., and at the time the plant was taken out of service in 1924 the total transmission pressure had attained about 18,000 v. The current was delivered at this voltage to a substation at Hanwell, a distance of about five and a half miles.

The relative positions held by the rival methods of supply in the eighties and nineties of the last century may be estimated from the nature of the current produced by the various undertakings then operating in the London area, as they are fairly representative of the variation of practice throughout the country generally. The following table gives particulars of the principal London stations in commission during the period in question, with the dates when they commenced supply. Allowance must, of course, be made for the relative importance of the undertakings. This may be judged by the equivalent number of 8 c.p. lamps connected to each system on 1 January 1892, when the approximate figures for the principal Companies were as follow:

Company	No. of lamps connected
Metropolitan Electric Supply Co., Ltd.	82,000
Westminster Electric Supply Corporation, Ltd.	62,000
Kensington and Knightsbridge Electric Lighting Co., Ltd.	38,000
St James and Pall Mall Electric Lighting Co., Ltd.	38,000
London Electric Supply Corporation, Ltd.	36,000
Chelsea Electricity Supply Co., Ltd.	28,000
Brompton and Kensington Electricity Supply Co., Ltd.	19,000
Notting Hill Electric Lighting Co., Ltd.	3,000

TABLE
Principal Electricity Supply Undertakings
in London. 1887 to 1897

Continuous current

	Ceased generating
Charing Cross Electricity Supply Co., Ltd.	
Maiden Lane (1887), D.C. three-wire 200/100 v.	1902
Lambeth (1896), D.C. 1,000 v.	1909
Chelsea Electricity Supply Co., Ltd.	
Draycott Place (1889), D.C. high tension	1898
Flood Street (1894), D.C. high tension	1928
Kensington and Knightsbridge Electric Lighting Co., Ltd.	
Kensington Court (1887), D.C. two wire, 100 v.	
Cheval Place (1890), D.C. two wire, 100 v.	1923
Metropolitan Electric Supply Co., Ltd.	
Whitehall Court (1888), D.C. two wire, 110 v.	
Notting Hill Electric Lighting Co., Ltd.	
Bulmer Place (1891), D.C. three wire, 240/120 v.	
St James' and Pall Mall Electric Lighting Co., Ltd.	
Mason's Yard (1889), D.C. three wire, 214/107 v.	1910
Carnaby Street (1893), D.C. three wire, 214/107 v.	1923
St Pancras Vestry	
Stanhope Street (1891), D.C. three wire, 240/120 v.	
King's Road (1895), D.C. three wire, 240/120 v.	
Shoreditch Vestry	
Coronet Street (1897), D.C. 1,100 v.	
Westminster Electric Supply Corporation, Ltd.	
St John's Wharf (1890), D.C. three wire, 220/110 v.	1910
Eccleston Place (1891), D.C. three wire, 220/110 v.	1922
Davies Street (1891), D.C. three wire, 220/110 v.	1921

Alternating current

Brompton and Kensington Electricity Supply Co., Ltd.	
Richmond Road (1889), single phase, 2,000 v. 83 cycles	1928
City of London Electric Lighting Co., Ltd.	
Bankside (1891), single phase, 2,000 v. 100 cycles	
County of London Electric Lighting Co., Ltd.	
Wandsworth (1897), single phase, 2,000 v. 100 cycles	
City Road (1897), single phase, 2,000 v. 100 cycles	
Hampstead Vestry	
Stoneyard (1894), single phase, 2,000 v. 90 cycles	
Islington Vestry	
Eden Grove (1896), single phase, 2,000 v. 50 cycles	
London Electric Supply Corporation	
Deptford (1889), single phase, 10,000 v. 83 cycles	
Metropolitan Electric Supply Co., Ltd.	
Rathbone Place (1889), single phase, 1,000 v. 100 cycles	
Sardinia Street (1889), single phase, 1,000 v. 100 cycles	
Manchester Square (1890), single phase, 1,000 v. 100 cycles	
Amberley Road (1893), single phase, 1,000 v. 100 cycles	1926

(*Note*: Sardinia Street was converted to D.C. 220 v. in 1897–1901. Manchester Square was converted to D.C. 220 v. in 1900.)

The above table brings to light the differences of practice that existed between the representatives of both the alternating-current and the continuous-current group of stations. In the case of alternating current stations the frequencies stated must be taken as approximate only. The practice of running each alternator on its own circuit, independently of the other machines, made questions of frequency of little importance, so that it was the custom to operate each engine or turbine at its most favourable speed, and let the frequency take care of itself. In the continuous-current group the Chelsea Company generated current at high tension and reduced the pressure to 100 v. for the consumers by means of batteries or continuous-current transformers in substations. The Shoreditch Vestry adopted the same system of generation, but Companies like the Westminster and the St James and Pall Mall employed low-tension current throughout, but adopted the three-wire method of distribution to reduce the losses in the mains. The Charing Cross Company had some of both systems. The alternating-current stations generated single-phase current only, for three-phase generation and transmission was not introduced into the London area before the joint station of the Kensington and Notting Hill Companies at Wood Lane was put into commission in 1900. The Paddington Station of the Great Western Railway Co. (1886) certainly had two-phase alternators, but as each phase supplied a separate and distinct circuit the machines were virtually single-phase generators. A genuine two-phase supply was, however, given from Willesden in 1899, this station having been erected to take over the load from the smaller generating stations of the Metropolitan Company at Rathbone Place, Sardinia Street, Manchester Square and Amberley Road.

One of the difficulties with which alternating-current stations were faced in the early days was that of operating the generators in parallel. It had been shown by Hopkinson in 1884 that parallel working was theoretically possible, and that the machines, once paralleled, would tend to keep in phase with each other. By 1887 devices for synchronizing alternators were available, and by the next year the parallel running of such machines was a matter of routine in several stations in the United States, incoming machines being brought into synchronism by means of a lamp before being switched into circuit. In this country, owing largely to the ill-omened Electric Lighting Act of 1882, there were in the eighties very few stations in which the parallel running of alternators could have been attempted. It had undoubtedly been tried at an early date, apparently with unfavourable results, for Mr J. E. H. Gordon, speaking in 1888, after two years' experience of the operation of his own large alternators at Paddington, and possibly on account of that experience, is reported as saying that:

Many people had tried to work alternating-current machines in parallel, but they did not work together till they had jumped for three or four minutes. Now, three or four minutes' jumping in the big machines might take a month's life out of 20,000 lamps, and that loss was rather a serious difficulty. Therefore they might take it in practice —because he was not talking of what might be done in the laboratory —that they would not couple alternating-current machines, and they must have very big machines if they were to get alternating currents to do the work.

It is difficult to say with precision when the practice of running alternators in parallel started in England, or to what station the credit should properly be given, but it is certain that the practice existed at Bournemouth in 1889, and was indeed obligatory when the load exceeded the capacity of a single generating unit, for there was only one high-tension cable to supply the town. Alternators were also being run in parallel in the Richmond Road Station of the Brompton and Kensington Electricity Supply Co. the same year, when the procedure was considered of sufficient novelty to warrant a visit of inspection by Major Marindin and Major Cardew of the Board of Trade. It is also on record that, in 1890, the Grosvenor Gallery Station had been operated in parallel with the Deptford Station, eight miles away, but this was not by any means the general method of working.

Many difficulties were encountered in the first attempts at parallel running, and Gordon was not alone in his belief that the practice was, to say the least, inadvisable. Various explanations of the trouble were put forward. Many engineers held that success could only be hoped for with machines of the Ferranti or Mordey types, which had no iron in their armatures, the argument being that the iron in an armature could not respond quickly enough to the high frequencies of alternation which were then usual. Hence there arose great debate as to the respective advantages of "copper-type" and "iron-type" machines, the discussion lasting several years. Other engineers believed that the transmission of power to the alternator by means of belts or ropes, in order to have a certain elasticity in the drive, was one of the essentials of success. Both at Bournemouth and at Brompton, where the machines ran satisfactorily in parallel, rope-driven copper-type alternators were employed, those at Bournemouth being designed by Mordey, and those at Brompton being of the Elwell-Parker design. At Paddington, on the other hand, the Gordon alternators were of the iron type running at a slow speed and directly coupled to tandem compound engines, of which the turning moment must have been particularly bad. It is hardly surprising that stations equipped with units of such a kind should experience difficulty with parallel running, especially as some of the alternators produced an E.M.F. with not the slightest resemblance to a sine curve, except that it was of a periodic nature. Examples of such curves of E.M.F. taken from actual machines built by leading Continental makers, are to be found in the 1900 edition of Professor S. P. Thompson's work on *Polyphase Electric Currents*, where they are given without comment, as representing acceptable practice.

To convince central station engineers that alternators could be operated successfully in parallel when driven directly by engines without ropes or belts, Mordey arranged a demonstration at the Thames Ditton Works of Messrs Willans and Robinson in July 1891. He brought down two of his own alternators, each having a rated capacity of about 35 K.W. at 666 R.P.M. and caused them to be coupled respectively to two Willans two-crank engines designed for a speed of 472 R.P.M. The engines were of standard type, and were taken directly from stock. The sets were run up to speed, and the alternators then synchronized by means of a transformer and lamp before being switched into parallel. As might be expected, they ran perfectly together under all conditions of load. One set was then caused to drive the other as a motor, and various other experiments were tried, all destined to confound those who denied that direct-coupled sets would work properly in parallel. The stubbornness of the critics is shown by the fact that one prominent engineer held that the tests had not been conclusive because the machines could be pulled out of step when paralleled through a resistance capable of absorbing the whole output of one of them! The result of the demonstration was to remove from the minds of all unprejudiced engineers all reasonable doubt as to the practicability of parallel running of direct-coupled sets, providing that both engines and alternators were of suitable design. By 1894 parallel running of alternators was common, though difficulties were still encountered. Rope-driven Mordey alternators, which had proved so successful at Bournemouth and elsewhere, when driven by steam engines, failed to give satisfaction when gas engines were tried as prime movers. The Coatbridge Central Station of the Scottish House-to-House Electric Lighting Co., was inaugurated in March 1894 with three Dick-Kerr horizontal double-acting side-by-side producer gas engines running at 180 R.P.M. and driving three 50 K.W. Mordey alternators at 600 R.P.M. by means of ropes. Single-phase current was generated at 100 cycles and 2,000 v. Although there were 360 explosive impulses per minute, and each engine had three flywheels, one on each end of the shaft and one between the cylinders, the latter flywheel carrying the driving ropes, satisfactory running in parallel could not be obtained, and for this reason it was decided in 1895 to replace the gas engines and producers by steam engines and boilers.

It is reasonable to suppose that the main cause of trouble in running these early alternators in parallel, was the uneven turning moment of the engines. This was of less importance when rope driving was employed, because of the elasticity of the ropes. With slow-speed alternators mounted on, or coupled directly to, the engine crankshafts, there was no such elasticity of the drive, and cyclic variations could only be smoothed out by the flywheel effect of the alternator. There was generally a small residual cyclic variation, enough to impair the parallel operation of the machine, and when parallel running became the rule in all stations it was the custom to mitigate the effects of the cyclic irregularity of slow-speed engine-driven sets, by paralleling them at the moment when the cranks of the incoming machine were in the same rotational position as those of the machines on the load. There was then no tendency for the alternators to be pulled out of step, as the cyclic variations of speed were simultaneous in all units. The identity of the angular positions of the crank shafts was secured by noting the position of the corresponding arms of the alternators, and only closing the switch when the correct relationship was attained. To enable this to be done, corresponding arms were marked with white paint so as to be readily distinguishable. The light from the station arc lamps produced a stroboscopic effect on the rotating machinery, permitting the relative positions of the white marks to be easily observed. When the latter had come into phase, and synchronism of the frequency had also been obtained, the machines could be put into parallel with certainty of satisfactory behaviour. An alternative method of ensuring that the cranks of the engines should be in the same angular position when paralleling was practised at the Willesden Station of the Metropolitan Electric Supply Co., Ltd., and possibly elsewhere. The indicator cock on the top end of the high-pressure cylinder of one of the engines in service was opened, and the corresponding cock on the engine of the incoming machine was also opened. The consequence was that each engine emitted a sharp puff of steam once in every revolution, and it was easy to tell, either by the eye or the ear, when the two puffs became coincident. The operator knew that the cranks were then in the same phase and the machines could be safely paralleled.

In many of the early alternating-current stations, such as Paddington (1886), Sardinia Street (1889), Forth Banks (1890), Manchester Square (1890), Amberley Road (1891), Cambridge (1892) and Scarborough (1893), there was no provision for paralleling the machines at all, as the designers of the stations did not believe this method of operation to be either necessary or desirable. The alternators at Amberley Road were paralleled experimentally in 1893, but with no intention of continuing the practice. It was held to be safer to keep each machine to its own feeder, or group of feeders, even if it could be shown that the alternators in the station were of such a nature that parallel operation was possible. There was a feeling that the failure of one machine, or even of one of the exciters, might cause a shut-down of the whole plant, a fear that seemed to have some justification from experiences at the Bankside Station of the City of London Company. In November 1893 the supply from this station was totally interrupted because one of the alternators lost its field and the fuse which should have isolated it from the bus-bars failed to do so. A 2,400 v. arc resulted, short-circuiting, of course, the other machines on the load, whose fuses consequently blew, one after the other. A similar accident, due to an exciter field going to earth, happened in the same station about a year later, causing a total extinction of lights in the offices and streets of the City at a time when everything was enveloped in a thick London fog.

In 1894 the "Battle of the Systems" was at its height. The relative position of the opposing forces may be estimated from the fact that at the beginning of that year there were about 373,000 lamps served by continuous current, as against some 320,000 served by alternating current in the London area. In the Provinces the corresponding figures were about 235,000 and 181,000 respectively. The great provincial cities of Manchester, Liverpool, Birmingham, Glasgow and Edinburgh were all committed to direct-current systems, while alternating current found its field chiefly in places where the population was less dense, and the consumers were scattered over a larger area. The efficiency of generation and distribution by the rival systems in those days will be gathered from a paper on "The Cost of Electrical Energy", by Colonel Crompton, read before the Institution of Electrical Engineers, and to be found in vol. xxiii of their *Proceedings*. In this paper comparative data were given concerning the operating results of twenty-three different undertakings, whence it was shown that the average heat consumption of the alternating-current systems of the Leeds, Bournemouth, Brompton, Newcastle-on-Tyne, Metropolitan and Eastbourne Companies was 161,000 B.T.U. per K.W.H. generated, and 245,000 B.T.U. per K.W.H. sold. In all these stations, the generating sets consisted of rope-driven alternators. On the other hand, the continuous-current stations of the Liverpool, Charing Cross, Birmingham, Hove, St James' and Pall Mall, Bradford, Brighton, Kensington, Westminster, and Knightsbridge undertakings had an average heat consumption of only 112,000 B.T.U. per K.W.H. generated, and 133,000 B.T.U. per K.W.H. sold. Some stations could certainly do a good deal better than these averages, for in the discussion on the paper, Professor A. B. W. Kennedy, the consulting engineer to the Westminster Company, stated that the Eccleston Place Station, with high-speed direct-coupled continuous-current generators and batteries, had shown a coal consumption of only 4·25 lb. per K.W.H. generated over a three months' period during the winter.

A few months after Crompton's paper, Professor S. P. Thompson took up the cudgels on behalf of alternating current in a paper he read at the 1894 Meeting of the British Association in Oxford. He pointed out that it had been demonstrated by Wilde in 1868 and again by Hopkinson in 1879 that two alternating-current machines could be run together synchronously, one as a generator and the other as a motor. He remarked on the recently invented Scott system of connecting the windings of a transformer so that three-phase current could be obtained from a two-phase source, and vice versa, and explained how over-excited synchronous motors could be employed as condensers to improve the power factor of alternating-current circuits. Finally he prophesied that "With such advantages in respect of motive power over continuous-current working, it can hardly be doubted that, save in a few special cases, the vast majority of Central Stations will henceforth be operated by alternating currents".

To this the supporters of direct current retorted that, while it might be true that the majority of new stations would generate alternating current, this was only because such stations would only be built to serve areas of little importance and with a low density of population, for all the great centres like Manchester,

Liverpool, Glasgow, Birmingham, etc., and the greater part of London were already committed to a direct-current supply. The advantages of alternating current, particularly as regards transmission, were, however, too obvious to be denied, and some engineers thought that a combination of the two systems was likely to prove the ultimate solution of the problem. An interesting example of a combined system was put into operation at Brighton in 1894. The town was then supplied, for the most part, by a three-wire direct-current system at 115 v. To supplement this, two motor generators were installed in the power station for the delivery of high-tension alternating current to a ring main laid through certain outlying districts. This main passed through two transformer substations, in which the secondary windings of the transformers could be switched on to the three-wire network. It was thus possible to feed the network either with direct or alternating current at will. The normal direct-current supply was given from about 11 p.m. until sunset the next day, when the network was changed over to the alternating-current system in order to take advantage of the lower transmission losses during the hours of heavy demand. About a hundred consumers, with an aggregate of some 4,000 lamps, were supplied on this dual system, which operated with complete satisfaction for several years, except for the momentary blinking of the lights at the times of changing over. The alternating current was measured to the consumers by the same meters as were used for the direct current.

The Battle of the Systems was a long contest, and though soon after 1900 it had become clear that the future of the industry was likely to lie with large power stations transmitting three-phase alternating current at high voltage to a number of conveniently situated substations, this still left open the question as to the form in which current should be supplied to the consumer. The advantages of the static transformer as compared with the running machinery necessary to give a direct-current supply, tipped the scales in favour of alternating current, and further weight was added by the development of the hardy and simple induction motor. There was, however, never any definite and complete defeat of the continuous-current system, comparable with that of the reciprocating engine by the turbine, or the tank boiler by the water tube boiler in power stations. It had to yield in scattered areas to the simplicity and convenience of the static transformer, but where the load was sufficiently dense it failed to hold its position, more, perhaps, on account of the desire for standardization than for any technical reason.

It may be that the Battle of the Systems is not yet over. In 1907 Lord Kelvin declared, "I have never swerved from the opinion that the right system for long distance transmission of power by electricity is the continuous-current system", and the development of the mercury-vapour rectifier may some day justify Lord Rayleigh's prophecy of many years earlier, that direct current would have its revenge in the final encounter.

CHAPTER VIII

GAS ENGINES AS PRIME MOVERS
IN POWER STATIONS

This Power hath worked no good to aught that lives.

TENNYSON

IT was at the York meeting of the British Association in 1881 that Sir Frederick Bramwell uttered his famous prophecy to the effect that in fifty years' time steam machinery would be found only in museums, its place being taken by the gas engine. When he said this there was probably not a gas engine in existence with an output of more than 12 H.P., and no engine had ever been worked with producer gas except one of 3 H.P. shown by Mr J. Emerson Dowson at the meeting in question. So sure was Bramwell of the truth of his convictions that he left in his will a sum of money to be awarded to a lecturer who should review the field of power production in 1931, being no doubt confident that such a survey would prove that his forecast had been fulfilled. The lecture was duly delivered by the late Sir Alfred Ewing, but the facts that he had to relate were hardly such as Bramwell had anticipated. Enormous as is the aggregate power of internal combustion engines, the gas engine accounts for no more than an insignificant proportion, and steam still reigns supreme in all stationary power plants, except for units of the very smallest sizes.

In spite of the great expectations which certain people held concerning the future of the gas engine at the time when the public supply of electricity was in the earliest stages of its development, it was very rarely that any of the power stations then established were provided with gas engines as prime movers. The first public power station to be so equipped was one erected in 1892 at Morecambe, in Lancashire, by the Morecambe Electric Light and Power Co. It started with three "Stockport" gas engines, each of 25 H.P. capacity, belted individually to three two-pole dynamos, which worked in conjunction with a 72-cell battery of accumulators. The station went into service on town gas, but was almost immediately rendered independent of the local gas works by the installation of two Dowson producers and the necessary auxiliary plant. The engines were of the single-cylinder horizontal type, with a trunk piston and two flywheels. They ran at 180 R.P.M. and transmitted their power to their dynamos by means of woven cotton belts which were found not to come off the pulleys so readily as ordinary leather belts. In a short time the plant was increased by the addition of a Crossley engine of 100 H.P. and a Dowson producer of corresponding size. The dynamo driven by the Crossley engine was a six-pole machine of the double-current design, capable of supplying either alternating or direct current, or both kinds simultaneously. Like the other dynamos it was driven by a cotton belt from the engine.

The gas from the producers passed first through a cooler consisting of a number of vertical cast-iron pipes, then, by way of an hydraulic box and water trap to a sawdust scrubber, and finally through a coke scrubber to the gas-holder. The sawdust in the first scrubber was renewed every three weeks, and the coke in the other scrubber every six months. The gas-holder served more as an equalizer of pressure than as a reservoir of gas, for its capacity was only sufficient to supply the engines for ten minutes when the station was working at its full output. The producers, therefore, had to respond almost immediately to any increase or decrease in the load on the station. They were supplied with steam at 60 lb. pressure from two small vertical boilers, which consumed about 120 lb. of coke per hour.

All the engines depended on hot-tube ignition. The ignition tubes were short lengths of wrought iron steam pipe, plugged at one end, and they had a working life of about 50 hr. only. Each engine had its own separate exhaust pipe carried upwards through the roof and terminated by a silencer consisting of an iron cylinder full of pebbles. The three smaller engines were started by using their dynamos as motors, current for this purpose being taken either from the battery or from other units that might be running at the time. The large engine was fitted with a Crossley "explosion starter". To set the engine in motion, the piston was put at the mid-point of the impulse stroke, and the roller of the exhaust cam lever made to ride upon a part of the cam which would cause the exhaust valve to open during every inward stroke of the piston. The ignition tube was well heated, and under these conditions an explosive mixture of gas and air was caused to flow through the part of the cylinder behind the piston by opening the cocks of the starter. In due course the charge exploded and the engine started. It was allowed to make six explosions in this manner, after which the speed was sufficiently high for the exhaust valve cam to be restored to its proper position, and the starter put out of action, when the engine would continue to operate in the normal manner. Regulation of speed was first effected by a combination of throttle and cut-out governing, but after some experience all governing arrangements were discarded. The engines were worked at full load all the time they were in service, their excess power, if any, being absorbed by the accumulators with which the dynamos ran in parallel.

With anthracite for the producers at 18s. per ton, the fuel cost per unit sent out from the Morecambe Station amounted to about 0·5 pence. A charge of 7d. per K.W.H. was made for the current taken by the few motors on the system, but current for lighting was supplied at a contract rate of £1 per annum for the eouivalent of every 16 c.p. lamp connected to the mains.

Before the station had been running very long, complaints began to be made of the nuisance caused by the gas and fumes from the producers, and as these were endorsed by the Medical Health Officer, the Municipal Council threatened to take action against the Company, who came to the conclusion in 1894, that the only way out of the difficulty was to remove the works to a new site altogether, further from the residential district. This scheme was, however, superseded by an agreement in 1896 for the sale of the whole undertaking to the Council for £3,897, an amount which was afterwards reduced to £2,718, although the Company had spent about £9,000 in establishing the business. The career of the station as a gas-driven plant was shortly afterwards brought to a close, the producers and gas engines being replaced by steam machinery.

The next gas-driven station was that of the Scottish House-to-House Electricity Co., Ltd., which went into commission at Coatbridge, Lanarkshire, in March 1894. The engine-room equipment, as originally installed, comprised three horizontal, two-cylinder, side-by-side, double-acting gas engines working on the four stroke cycle. They were of the "Kilmarnock" type, constructed by Messrs Dick, Kerr and Co., Ltd., and had a rated capacity of 120 I.H.P. at 180 R.P.M. Each engine drove a 50 K.W. Mordey alternator at 600 R.P.M. by means of eight one-inch ropes from a grooved pulley on the centre of the crankshaft. Two other flywheels, one at each end of the crankshaft, were provided to equalize the turning moment. The alternators generated single-phase current at 2,000 v. and a frequency of 100 cycles. The engines were started by steam. A small pipe from the boiler led to a steam chamber at the back of one of the cylinders, and this chamber could be put into communication with the combustion space by means of a valve controlled by a hand lever. By the manipulation of this lever, the operator could admit puffs of steam behind the piston at appropriate moments, and could thereby run the engine up to a speed sufficient for it to operate on gas in the ordinary way. The producer-house, adjoining the engine-house, contained two gas producers of a modified Dowson type, with arrangements for removing the clinker formed when working. Anthracite was used as fuel, being gasified by a blast of mixed air and steam. The necessary steam was raised in a vertical coke-fired boiler. After leaving the producer, the gas passed through a cast-iron cooling stack, a hydraulic dust box, a coke scrubber and a sawdust filter to the gas-holder, the latter being 15 ft. in diameter and 10 ft. in height. Each of the producers was capable of making sufficient gas for the development of 120 B.H.P.

The Coatbridge plant gave trouble from the start. The engines proved incapable of developing the required power with the quality of gas supplied to them, and were replaced almost at once by others of the same make and type, but rated at 140 I.H.P. There was also great difficulty in running the alternators in parallel, in spite of the elasticity of the drive and the heavy fly-

PLATE XXII. COATBRIDGE POWER STATION, 1894

From contemporary engraving in *The Electrical Review*

wheel equipment of the engines. Tube ignition was employed on the engines, and this was also a source of trouble and expense, for the tubes, which were of wrought iron, would only last for a day, and each engine had four of them in use all the time. In consequence of the various troubles attending the operation of the plant, its high working costs, and particularly the unsatisfactory running of the alternators in parallel it was decided in 1895 to replace the whole of the gas equipment by steam machinery. At the end of June in that year, the gas plant was shut down, the supply being taken over by three 100 H.P. Brush compound non-condensing engines of the marine type, taking steam from a pair of 26 by 7 ft. Lancashire boilers fitted with Vicars' stokers. Under these conditions the station consumed 35·5 lb. of coal per K.W.H. as compared with 26·4 lb. when the gas plant was in service, but as coal suitable for the boilers could be bought at about a quarter the price of anthracite, the use of the steam plant resulted in a substantial saving in fuel costs. The conversion of the station was a confession of defeat but it must be remembered that Coatbridge was the first alternating-current station to attempt to operate with gas engines, and even in steam stations parallel running was still regarded as a procedure requiring special characteristics in the machinery.

Another early gas-power station was that put down by the Midland Railway Company at Leicester in 1894 for the lighting of the Railway Station and Goods Yards. The Company possessed steam-driven stations in six other centres at the time, but determined to try a gas station at Leicester as an experiment. The plant consisted of two Dowson producers working on anthracite, and supplying gas to six single cylinder horizontal Crossley engines, with porcelain tube ignition. There were no batteries in the station. Three of the engines drove 60-light Brush arc-lighting machines, and the other three drove direct-current dynamos for incandescent lighting at 110 v. All engines were belted to their respective dynamos. The connected load comprised 141 arc lamps and 288 incandescent lamps, the output of the station being about 274,000 K.W.H. per annum. The plant ran for many years, and gave satisfaction.

In 1895 another gas-driven station was put into commission, this time in the City of Belfast. It was a Municipal enterprise, undertaken to supply continuous current on the three-wire system, using town gas to drive the engines. One of the objects in establishing the station was to relieve the demand on the Municipal gas works, as more light could be obtained by using the gas to make electricity than by burning it in the usual way. The station was designed by Professor A. B. W. Kennedy, and erected on a site between Chapel Lane and Marquis Street, not far from the centre of the City. It went into service on 23 January 1895, with an equipment of six generating sets, each consisting of a horizontal Dick-Kerr gas engine driving a dynamo by means of ropes. The engines were of two sizes, those of the four larger units having each two double-acting cylinders, 13·75 and 13·5 in. diameter respectively by 20 in. stroke. They had a rated capacity of 120 I.H.P. at 160 R.P.M. In contrast to the practice of the same makers with regard to the Coatbridge engines, those at Belfast were built as tandem single-crank machines. On one end of the crankshaft was a 37 cwt. flywheel, 8 ft. 5 in. in diameter, and on the other side was a 29 cwt. rope pulley of the same size, this end of the shaft being carried in an outboard bearing. A 60 K.W. bipolar shunt-wound Siemens dynamo was driven from the pulley by a set of eight cotton ropes. The dynamos were placed behind their engines, so that the rope drives could be accommodated in the space alongside the cylinders, the layout being thus as compact as possible. It necessitated, however, the running of the engines in what would generally be regarded as the backward direction, that is, with the crank moving upwards instead of downwards when passing through its outer dead centre. This had the result of keeping the slack of the ropes on the upper side of the drive, an obviously desirable arrangement. The dynamos generated current at 220–240 v., according to the regulation of their fields.

The two smaller sets were of the same general type, but had single-cylinder double-acting engines. The cylinder diameter was 13·5 in. and the stroke 20 in. These engines developed 60 I.H.P. at 160 R.P.M. and their dynamos 26·5 K.W. at 750 R.P.M. The sets could be used either as main generating units, or as balancers across the three-wire system, and to this end their armatures were provided with double windings. Each winding developed from 110 to 120 V., and had its own commutator. The windings could be connected either in series or in parallel by the operator on the switchboard, according as the machines were wanted for generating or for balancing purposes.

The engines could be started either by motoring their dynamos from the batteries, or by means of air at 100 lb. pressure. They were fitted with the wrought iron tube ignition usual in those days. The tubes had an economical life of only about 15 working hours, as after that they became sluggish in firing the charge. Governing was effected entirely by throttling the charge, no explosions being cut out. This system was said to be entirely satisfactory, and it was claimed that the load could be varied from an overload of 20 % down to zero without the engines missing an explosion. The exhausts were taken by a common exhaust main to a silencer which consisted of a cylindrical chamber fitted with a number of perforated plates. A vertical pipe from the silencer discharged the gases above the roof. Each engine had its own circulating pump, which returned the cooling water to a 10,000 gallon overhead tank. A direct supply of cooling water was also available from the town mains.

The engines worked on ordinary coal gas enriched by carburetted water gas, as used for lighting purposes. The gas was charged for at the rate of 2s. 3d. per 1,000 cu. ft., the price to domestic consumers in Belfast being 2s. 9d. per 1,000 cu. ft. The supply was passed through a duplicated meter before going to the engines. The consumption of the large sets was stated to be 34·2 cu. ft. per K.W.H. at full load. During the period from the start of the plant on 23 January to 31 December 1895, the output was 82,771 K.W.H. for a consumption of about 46·8 cu. ft. per K.W.H. 87 % of these units were sold, 6·7 % were wasted in the batteries, 6·0 % were used for station lighting and 2·3 % were lost in the mains. The quantity of gas consumed per K.W.H. sold was 53·7 cu. ft., costing about 1·45d. The total generating expenses amounted to 2·82d. per K.W.H. of the power generated, and 3·24d. per

K.W.H. sold, figures which were not considered at all bad at the time.

The station included a battery-room 28 ft. sq., with two batteries, each of 68 E.P.S. cells, with a capacity of 500 amp.-hours on a 5-hr. discharge. The regulation of the battery voltage was not performed in the usual way by cutting out end-cells when the pressure had to be reduced, and restoring them to the circuit in the converse circumstances. It was accomplished, instead, by putting a number of cells at the middle-wire end of the positive battery in parallel with a number of other cells at the middle-wire end of the negative battery. This was done by a special arrangement of sliding contacts, operated from the switch-board. The other ends of the two batteries were connected through ammeters and switches to their respective bus-bars. The switch-board was of slate, with the machine and feeder bars mounted vertically on the front. Each of these bars had three holes, through which plugs could be inserted to make connection with either of the three horizontal bus-bars at the back of the board. The voltmeters used were Kelvin multicellular electrostatic instruments.

The equipment was increased in 1896 by the addition of two 150 K.W., 380 R.P.M. units, each consisting of a vertical four cylinder, two-crank, single-acting engine, built by the Acme Gas Engine Co. of Glasgow, and directly coupled to two dynamos of 75 K.W. capacity each. Each engine had two cylinders in tandem on a crank, the lower cylinder having a trunk piston and depending upon splash lubrication from the crank case.

The Belfast gas engine station appears to have run without giving rise to the complaints caused by the operation of other gas power plants of the same period, but whether the growth of the load caused concern about the gas supply, or whether the reason is to be sought amongst the vagaries of Municipal politics, the fact remains that it had not been running very long before the Council turned their thoughts to steam. Plans were therefore prepared in 1897 for a new steam station, which was erected in New Bridge Street, and formally opened on 18 October 1898. It started with three Lancashire boilers, 30 ft. long by 8 ft. in diameter, working at 180 lb. pressure, and three high speed Belliss engines directly coupled to Parker dynamos. Supply was given by a three-wire system with 440 v. across the outers, and two 500 amp.-hr. batteries were operated in conjunction with the generators. These were regulated by end-cells in the usual manner, because the Kennedy system of regulation employed in the older station was not a success on account of the irregular charging and discharging of the overlapping cells.

Amongst the several unfortunate attempts to operate gas power stations, must be reckoned that of the Leyton Urban District Council, whose station went into commission on 15 September 1896 with two Premier single-cylinder horizontal gas engines running at 200 R.P.M. and driving 42 K.W. E.C.C. dynamos at 700 R.P.M. by leather belts. Gas was provided by two Dowson producers burning anthracite. A battery of 85 Tudor cells, with a capacity of 1,100 amp.-hr. on a 10-hr. discharge, allowed the engines to be shut down during the periods of light load. Supply was given on the three-wire system with 300 v. across the outers. When the engines were not running, the battery was connected between the positive outer and the middle wire, with a motor generator across the other side of the system. The load on the undertaking grew rapidly, two more generating sets having to be put down and then another four, so that within two or three years there were eight gas engine sets and five producers in commission. The residents in the neighbourhood, however, soon began to complain of vibration and noise, and particularly of the back-firing to which the engines were prone. In February 1900 one of them obtained an injunction against the Council in the Courts, and as this did not result in an abatement of the nuisance, an Order of Sequestration was actually granted the next year. Meanwhile the Council had put down some steam plant, and their appeal against the Sequestration Order was only allowed on condition that they undertook not to run the gas engines any more. So ended the gas power station at Leyton, on the equipment of which some £9,741 had been spent.

Another attempt to give a supply of electricity from a gas-driven power station was made by the Municipality of King's

Lynn, which inaugurated such a supply in August 1899. The plant put down consisted of four Fielding and Platt producers and four gas engines by the same makers, each engine driving a 40 K.W. direct-current dynamo. A battery with a capacity of 400 amp.-hr. on a 10-hr. discharge worked in conjunction with the generators. The gas producers worked with steam at 30 lb. pressure supplied by a small vertical boiler, and each producer was guaranteed to deliver 13,000 cu. ft. of gas per hour at 1·5 in. water gauge. The gas was made from Welsh anthracite, and had a calorific value of 145 B.T.U. per cu. ft.

After leaving the producers the gas passed through a set of vertical cast-iron cooling pipes, and thence by way of an hydraulic seal, a coke scrubber and a sawdust scrubber, to a 3,000 cu. ft. gas holder, 20 ft. diameter by 10 ft. lift. The engines, which worked on the ordinary four-stroke cycle, were provided with a compressed air starting device, double tube ignition, and hit-and-miss governing. They were rated at 65 B.H.P. at 150 R.P.M. and were belted each to a 40 K.W. two pole overtype shunt-wound dynamo working at 450–500 v. Their fuel consumption at full load was stated to be 85 cu. ft. of gas per K.W.H. or the equivalent of one pound of anthracite.

In 1901, when extensions to the station became necessary, it was decided to proceed with steam plant, and a Belliss engine driving a Laurence Scott dynamo was installed, steam being provided by a Lancashire boiler. Consumers were supplied on the three-wire system, with 400 v. between the outers, the voltage across the two sides of the system being regulated by balancers.

A review of gas-driven power stations established before 1901 is little better than a record of failures. It is a monotonous story of the replacement of gas engines by direct coupled steam machinery, or of the adoption of such machinery when extensions had to be made. The early gas engines, however, were all made to drive their generators by ropes or belts, and although this alone is not sufficient to account for the difficulties of the stations, it was certainly a drawback to their success. More important were the troubles from vibration and noise, and the impossibility of a gas plant requiring expensive anthracite as fuel to compete economically with steam plant using cheap bituminous coal, in spite of having a higher thermal efficiency. The gas engine, nevertheless, still had its supporters who believed that with more suitable machinery, gas power would be found preferable to steam. Undeterred, therefore, by lack of success elsewhere, the Walthamstow Urban District Council decided to commit their fortunes to gas, and put down a producer gas station which was formally opened on 20 September 1901.

The Walthamstow Station went into service with three Westinghouse vertical three-cylinder gas engines, each direct coupled to a 66 K.W. multipolar dynamo. Power was supplied on the three-wire system with 460 V. across the outers. The engines worked on the Otto cycle, and were rated to develop 100 B.H.P. at 168 R.P.M. They were provided with low tension electrical ignition, with mechanical make-and-break contacts within the cylinders. The engines were started by compressed air at 190 lb. pressure, the reservoirs being charged by compressors driven by belting from the engines. A 40 ft. cooling tower, with fan draught, served to keep down the temperature of the cooling water for the engines. Gas was made from anthracite in two Dowson producers, equipped with the usual coke and sawdust scrubbers, etc., the steam for the producers being supplied by two small vertical boilers working at 40 lb. pressure. The purified gas passed into a holder 25 ft. diameter by 12 ft. deep, and thence into a 20 in. main from which a 5 in. pipe branched off to each engine. The station contained a battery of 284 Tudor cells with a capacity of 300 amp.-hr. on a 1-hr. discharge. A booster raised the line pressure for battery charging, and a balancer was provided for regulation of the voltage across the two sides of the three-wire system.

By the end of 1907 the equipment of the station had been increased to eight Dowson producers and thirteen gas engines. The latter were all of the Westinghouse three-cylinder vertical type, and included four of 100 B.H.P., three of 125 B.H.P., and six of 250 B.H.P., making a total of 2,275 B.H.P. of gas driven machinery. The battery had also been enlarged to a capacity of 1,000 amp.-

hr. In spite of the success which might be inferred from these enlargements of the gas plant, it was decided in 1907 to adopt steam machinery for the extensions then determined on, and steam plant of the normal type was installed, the first units consisting of two 300 K.W. Belliss-Siemens sets with Ledward-Beckett condensers working in conjunction with three Balcke cooling towers. Steam was supplied by a pair of Babcock and Wilcox boilers.

CHAPTER IX

RECIPROCATING ENGINES IN
CENTRAL STATIONS

I have touched the highest point of all my greatness,
And from that full meridian of my glory
I haste now to my setting.

SHAKESPEARE

THE large slow-speed generating unit never attained the same popularity in Great Britain as on the Continent and in the United States. When the semi-portable engines belted to their generators, which were characteristic of the earliest installations, gave way to engines of the stationary type, belt or rope driving was still employed in order to retain the desired high speed of the generator. There were, of course, many exceptions to this rule, among them being the Gordon units at Paddington, but generally speaking, rope-driving held its own until the appearance of the Willans central valve engine in the eighties. The high rotational speed of this engine made it particularly suitable for the direct driving of generators, and this, combined with the reliability due to its mechanical perfection, soon gained for it an overwhelming predominance in central station work. It had, indeed, no real competitor at all for several years, until the Belliss high-speed engine with its advantages of double-acting cylinders and forced lubrication, entered the same field. Even the latter type, however, was not well suited for outputs much in excess of 1,500 K.W., and it is probable that with the demand for larger powers, central station engineers would have had to fall back on the slow speed unit, had not reciprocating engines of every type been rendered obsolete, so far as main generating sets were concerned, by the introduction of the Parsons steam turbine.

The position as regards prime movers at the end of 1891 was summarized in a leading article in the *Electrical Review* published on 22 January 1892. It was then stated that the aggregate capacity of the plant at work in British Central Stations possessing

installations of 300 H.P. and over, was just under 33,000 H.P. Of
this machinery 22,300 H.P. or 68 % of the whole, consisted of
Willans engines direct coupled to their generators. Of the remainder, 4,800 H.P. or 14 %, was accounted for by the slow speed
rope-driven sets in the Deptford Station, and about 2,600 H.P. or
8 %, by the belted units in the Sardinia Street Station which was
of American design. The Parsons turbo-alternators at Newcastle
contributed only 500 H.P. or about 1·5 %, leaving no more than
about 2,800 H.P. for all other kinds of engines put together. The
supremacy of the Willans engine was, in fact, virtually un-
questioned, then and for some years afterwards. In 1895 it was
stated by Mr Mark Robinson, that the total engine power in
British Central Stations amounted to some 101,390 I.H.P. of
which 53,340 I.H.P. was provided by Willans engines, and more
than 4,000 I.H.P. by high speed engines of other makers. On the
same occasion, a total of 31,000 I.H.P. was claimed by Mr J. S.
Raworth for the engines of the Brush Co. which were then of the
"launch" or small marine type, running at a fairly high speed,
and usually driving their generators by means of ropes. The late
Mr Frank Bailey, who was at one time responsible for the
operation of 35 Willans engines simultaneously, said in his
anniversary address to the Electrical Power Engineers Association
at Manchester in 1931:

The perfection of the engine and its high speed of 350 R.P.M. un-
fortunately restricted competition, and for some years there was a
monotonous repetition of small generating works all over the country
with the magic formula "Willans engines : Babcock boilers" which
satisfied most conditions and assisted the dynamo maker who sent his
machine to the Willans works for assembly on the combined bed-
plate, and had it tested there before delivery. Although these engines
were gradually increased in size, it is interesting to note that in 1894
at Bankside, two 400 K.W. alternators were each driven by two sets of
Willans engines, one at each end of the alternator, a magnetic clutch
coupling up the second engine at about half load. After this date, still
larger engines of this type were constructed, but the results were not
so good.

It may be added that a Willans engine of 2,500 I.H.P. was shown
at the Paris Exhibition of 1900, but such sizes never came into

general use. Further testimony to the popularity of the Willans engine is afforded by the fact that in 1903 the Westminster Electric Supply Corporation had altogether 49 Willans engines of an aggregate capacity of 9,330 K.W. in its three generating stations. The small average output of the units may seem remarkable, but Professor A. B. W. Kennedy, the consulting engineer to the company, believed in small units. In 1891 he is reported to have said that "the best unit for a central station should not exceed 200 H.P. for the reason that with engines of larger power it was extremely difficult to keep them fully loaded, while the efficiency of a steam engine of 2,000 H.P. is but little better than one of 200 H.P." Curious as such an opinion may seem nowadays, Professor Kennedy was by no means alone in his views, for in 1904 the electricity supply of Liverpool was furnished by no less than 78 Willans and Browett engines with a total capacity of 24,825 K.W. and an average size, therefore, of less than 320 K.W. apiece.

It is interesting to note the impression made on a competent foreign observer by British Power Station practice in 1894. In November of that year an article by Dr Guido Semenza, one of the most distinguished electrical engineers on the Continent, appeared in *L'Elettricista* giving the author's views of the electrical situation in London. In the course of the article, which was translated in *The Electrician*, Dr Semenza remarked:

While, for example, they are reluctant to construct multi-polar dynamos, they have on the other hand abandoned the slow speed steam engine and transmission by belt and rope gearing, and have adopted, in the Willans engine and the principle of direct driving, two novelties which are slow to make their way in other countries. They use, with great facility, alternating currents for incandescent lighting, but they still look upon the distribution of power by polyphase currents in the same light as we do the dirigible balloon.

These prejudices come out strikingly on a visit to one of the London Central Electricity Stations. Let us visit one at random, no matter which, as they are almost all of the same stamp. An unimposing entrance, a gloomy little back alley, and at the end the workshop of the station. A covered passage, broad enough to allow a coal cart to pass, traverses the workshop. On the left-hand side, we will say, is the coal store, and to the right up a few steps, we find the boiler room. I call it the boiler room for the sake of the word. There is a long row of Babcock and Wilcox boilers arranged along the passage with a few yards between them and the wall. Proceeding in the same direction, we enter the engine room. Along the wall, separating it from the boiler room, there are installed parallel to the boilers, half a dozen or more Willans engines coupled to as many stout bi-polar dynamos mounted on the same bed-plate. On the wall facing us is the switch-board. Into the engine-room a grey light falls from a skylight, the walls are grimy and between one machine and the next, or the wall, there is hardly room enough to pass.... To complete the description of one of these stations, I should say that there is always a double battery of accumulators; that the distribution is on the three-wire system, and that the capacity of the stations varies from 60,000 to 100,000 incandescent lamps.

They are not, however, all of this type. Some of them have original characteristics which should be pointed out. An entirely new station (the author evidently refers here to the Wandsworth Station of the County of London Electric Lighting Company) has just been started on the right bank of the Thames, under the direction of Mr Mordey of the Brush Company. The general arrangement is still the same, but there is more room, more grandeur, and even a little elegance. The generators are Mordey-Victoria alternators coupled direct to marine type engines running at about 100 R.P.M. These alternators, from the mechanical point of view, are much more impressive when actually seen than would be gathered from a written description. The delicate fixed armature is really very solid and rigid, and it seems that de-formations are rare. The whole arrangement of interrupters and switches for the 2,000 v. circuits is very ingenious and also new. The handles are connected to each other in such a way that the various connections can only be made in a certain order. This, and the fact of the whole being enclosed in steel columns, appears to safeguard against all possible dangers.

The difference between British and foreign practice in the matter of engines was undoubtedly due to the perfection of the high-speed direct-coupled engine, primarily by Willans, and later by Belliss and other makers. So long as units of not more than a few hundred kilowatts capacity could satisfy the needs of central stations, the high-speed engine fulfilled all requirements much more cheaply and efficiently than its slow-speed rival. The former attained about the limit of its power in the 1,560 K.W. units running at 184 R.P.M. installed in the Grove Road Station in

1902. Had the steam turbine not intervened, it is possible that engineers would have been driven to the employment of the slow speed reciprocator for the larger units required about this time, but the coming of the turbine changed the entire outlook. A pair of 1,000 K.W. turbo-alternators were running in 1900, and so rapid was the progress made with turbine machinery that the large slow speed engine was obsolete almost before it had gained a footing in British Power Station practice. For a few years, however, round about the beginning of the century, it made a brief bid for popularity. It had, of course, always been available, for there were innumerable makers of slow-speed engines for textile mills or ship propulsion who would have been only too pleased to supply their products to power stations, but such success as it had in entering this field was almost entirely due to the influence of foreign practice.

In 1899 three Allis engines of 1,000 K.W. capacity each were put down in the Bankside Station of the City of London Company, to be followed shortly by a series of six 2,000 K.W. Musgrave engines, all driving Westinghouse continuous current generators. The Willesden Station of the Metropolitan Company started in 1899 with three 1,500 K.W. Westinghouse engines and alternators, and went on to Sulzer engines of 3,000 K.W. capacity. The Bow Station of the Charing Cross Company commenced operations in 1902 with three 1,600 K.W. 83·3 R.P.M. Sulzer horizontal engines, and continued with a pair of 4,000 K.W. vertical engines by the same makers and running at the same speed. The Bloom Street Station of the Manchester Corporation, which started in 1901, contained four 1,800 K.W. Musgrave engines driving Westinghouse generators, and the Stuart Street Station, which came into operation the next year, had six 1,500 K.W. Yates and Thom engines, to which a pair of 4,000 K.W. Wallsend Slipway engines running at 75 R.P.M. were added in 1905. These were said to be the largest reciprocating engines in any power station in Europe. They were of the four-cylinder triple expansion vertical type with a H.P. cylinder 37 in. diameter, an I.P. cylinder 59 in. diameter, and two L.P. cylinders 72 in. diameter, all with a stroke of five feet. A flywheel alternator weighing

112 tons was carried on an extension of the crankshaft. They were fitted with Corliss valves, and worked with steam at 190 lb. pressure, superheated to 500° F. Each engine was rated at 6,000 I.H.P. but would carry 6,500 I.H.P. for two hours. When exhausting to atmosphere, the output was 5,000 I.H.P. An engine efficiency, that is, the ratio of B.H.P. to I.H.P., of 90% was guaranteed, and a steam consumption of 11 lb. per I.H.P. with a 27 in. vacuum, and 18 lb. when running non-condensing. The Pinkston Station of the Glasgow Corporation which was started in May 1901 was another plant equipped with large slow-speed engines, of which two were built by Messrs Musgrave and two by the Allis Company, all having a capacity of 2,500 K.W.

The last important installation of slow-speed engines in a British power station was that of the London County Council in their Greenwich Station which was opened on 26 May 1906 with four units rated at 3,500 K.W. each, but with an overload capacity of 4,375 K.W. The steam pressure was 180 lb. per sq. in., and the speed of the engines was 94 R.P.M. The engines, which were built by Messrs Musgrave and Sons, of Bolton, were of what was known as the "Manhattan" type, first used in the power station of the Interurban Rapid Transit Co. of New York. They carried a flywheel alternator at the centre of the crankshaft, and the overhung crankpin at each end was driven by two cylinders, one vertical and the other horizontal. The whole unit consisted really of two independent compound engines, one driving each end of the shaft. The American practice had been to make the H.P. cylinder horizontal and the L.P. cylinder vertical, but the converse arrangement was adopted at Greenwich with the idea of getting better drainage. The H.P. cylinders were 33·5 in., and the L.P. cylinders 66 in. diameter, all with a stroke of 4 ft. The valve gear was of the Corliss type, and each L.P. cylinder exhausted into its own surface condenser in the basement.

It was originally intended to complete the Greenwich Station by another four engines of the same kind, but these were never ordered. By 1910, only four years after its inauguration, the Station contained four 5,000 K.W. turbo-generators and by 1922 the last of the original slow-speed sets had been thrown on the

PLATE XXIII. GREENWICH POWER STATION, 1914

By the courtesy of H. W. Dickinson, Esq.

scrap heap and replaced by turbine machinery. The decision to install reciprocating sets at Greenwich in the first place is hardly one upon which the authorities or their engineers were to be congratulated. Nearly two years before the station was opened there were Parsons turbines running at Carville Station, on the Tyneside, with an economical rating of 4,000 K.W., an overload capacity of nearly 6,000 K.W. and a steam consumption of only 15·4 lb. per K.W.H. The huge Greenwich engines were thus out of date long before they were ever erected, and their installation was an ill-advised effort to avoid the inevitable. The reciprocating engine had, indeed, reached the limits of its development for power station purposes, and much as one may regret the disappearance of the great masterpieces of the engine builders, the time of their usefulness had come to an end.

CHAPTER X

THE INTRODUCTION OF THE STEAM TURBINE

And all that moves between the quiet poles
Shall be at my command.

CHRISTOPHER MARLOWE

THE first turbo-generator in the world was constructed by Sir Charles Parsons in 1884. It was a direct-current unit capable of developing about 75 amp. at 100 v. when running at 18,000 R.P.M. This historic machine, after many years of useful work, was properly given an honoured resting-place in the Science Museum at South Kensington, where it may be seen among other famous engineering relics. Its first appearance may be said to have coincided with the beginnings of the power station industry, on whose future it was destined to have such an overwhelming influence. Indeed, but for the invention of the steam turbine, the industry, as we know it to-day, could never have been created at all, for without the immense and efficient turbo-generating units, characteristic of modern power stations, electricity would have remained a costly luxury, instead of becoming one of the universal necessities of industrial and domestic welfare.

The steam turbine, it may be mentioned, was not the first embodiment of Parsons' efforts to provide the electrical industry with a better prime mover than the ordinary type of steam engine. During the years immediately preceding his invention of the turbine, Parsons was active in developing an epicycloidal engine, based on the geometrical principle that, if one circle be made to roll round the interior of another circle of twice the diameter, any point on the circumference of the first circle will oscillate along a straight line. His engine comprised four cylinders revolving round a central crankshaft at half the speed of the latter, which was directly coupled to the dynamo and drove it at several thousands of revolutions per minute. About forty of such engines were constructed by Messrs Kitson and Co. of Leeds between 1881 and 1883 while Parsons was associated with that firm. One of the earliest was used for driving an arc-lighting dynamo at the steel works of Messrs Cammell and Co. in Sheffield. According to a contemporary report, this installation was "a great success", largely on account of the "motor, which is a Parsons patent engine, made by Messrs Kitson and Co. of Leeds, and works the dynamo machine direct without the intervention of belting, thus ensuring a good speed combined with regularity."

The turbo-generator did not enter the power station field immediately. It found its first employment in ship-lighting work and in similarly isolated duties on land. By 1887 more than a hundred turbo-generators were in service, but so far no electric lighting Company had shown any particular interest in the invention, or, at any rate, none had thought sufficiently well of it to install a turbine in any power station. The credit of being the first to take this important step in the development of the industry belongs to the Newcastle and District Electric Light Co., Ltd., a Company registered on 14 January 1889 to give a supply of electricity in the area indicated by its name. Its Board of Directors included men with great experience of the kind of work they were undertaking, such as Lord Crawford, the Chairman of the London Electric Supply Corporation, who were then constructing their great generating station at Deptford, Mr A. Wade who was also connected with the same Corporation, and Sir Charles Parsons himself, who was at that time a partner in the firm of Clarke, Chapman, Parsons and Co., Ltd., of Gateshead, as managing Director. In anticipation of obtaining the necessary powers from the Board of Trade, which, it may be said, were granted in 1891 without opposition, the Company had secured a site for its projected station at Forth Banks, had ordered the machinery and had obtained sufficient promised consumers to justify the commencement of operations.

The Forth Banks Power Station went into commission in January 1890, with an initial equipment of two 75 K.W. Parsons turbo-alternators constructed by Messrs Clarke, Chapman, Parsons and Co. at Gateshead, and supplied with steam from three boilers of the Lancashire type, built by Messrs Hawthorne, Leslie and Co., Ltd. Properly to appreciate the enterprise of the Newcastle and District Electric Light Co., Ltd., in deciding to trust entirely to steam turbines, it must be remembered, not only that no steam turbine machinery of any kind had previously been employed in any public power station, but also that the units they ordered were of greater capacity than had yet been constructed, and were furthermore the first turbo-alternators to be built for any purpose. The sets ran at 4,800 R.P.M. and produced single-phase current at 1,000 v. and 80 cycles. Each had its own exciter on the end of the alternator shaft, as is common in present practice. The turbines worked with saturated steam at a pressure of 140 lb. per sq. in. and exhausted at atmospheric pressure.

In 1892 the Company decided to improve the economy of the turbines by operating them in conjunction with a condenser. Their exhaust steam was taken by a wrought iron galvanized pipe, 2 ft. in diameter, to a condenser having a cooling surface of 1,512 sq. ft., composed of 790 brass tubes, 9 ft. 8·5 in. long and 0·75 in. in diameter. The condenser and pumps were placed in a pit, above which a tandem compound pumping engine was supported by beams. The engine had cylinders 6·25 in. and 11·5 in. diameter by 9 in. stroke, and ran at 60 R.P.M. The air pump and circulating pumps were driven by wooden spear-rods from a rocking lever actuated by the crosshead of the engine. Nearer the fulcrum of this lever was attached a spear-rod of wrought iron piping, operating the ram of the cold-water pump which was 4 in. in diameter and had a stroke of 9·25 in. It served to deliver the condensate discharged from the air pump to an overhead tank in the boiler room. The air pump was of the two-stage reciprocating type with a stroke of 18 in. The circulating pump, which had a delivery of 550 gallons per min., had an equal length of stroke. It was designed to give a constant discharge with an intermittent suction. To obtain this result it was fitted with a bucket 14 in. in diameter, to which was connected a ram of 10 in. diameter. Water was drawn in under the bucket on the up-stroke, and during the down-stroke; this water was delivered to the top of the ram, but as the latter had only half the area of the bucket, half of the water had

PLATE XXIV. FORTH BANKS POWER STATION, 1892

to escape by flowing through the tubes of the condenser. During the following up-stroke, the remainder of the water was delivered by the ram, a fresh supply meanwhile coming in underneath the bucket. All valves were of the indiarubber disc type, and an airvessel of 11·25 cu. ft. capacity was placed on the suction pipe to steady the flow.

The turbines at the Forth Banks Power Station had been constructed on the axial-flow principle in accordance with Parsons' original patent, but before they could be put into service, differences of opinion arose between Parsons and his colleagues, with the result that the partnership was dissolved. He therefore decided to start manufacturing on his own account, and with the aid of friends he established in 1889 the firm of C. A. Parsons and Co., Ltd., with works at Heaton, on the outskirts of Newcastle. Under the terms of his agreement with Messrs Clarke, Chapman and Co., however, he was debarred from the use of the patents he had taken out whilst associated with them, so that he was unable for the time being to build turbines in which the steam passed axially through the machine in the natural way. His only means, therefore, of continuing his turbine work was to develop a type in which the flow of steam was not axial but radial, and with his usual resourcefulness and energy he proceeded immediately to devise a turbine of this kind.

Although handicapped by being forced to work on lines that he knew were not the best, he was so successful that in a very short time he was able once more to enter the power station field. Cambridge afforded him the desired opportunity. The Corporation of that town had acquired Electric Lighting Powers in 1890, but like many other municipalities, they did not care to embark upon the somewhat speculative business of electricity supply themselves. They therefore agreed to entrust their parliamentary powers for the lighting of the town to Messrs C. A. Parsons and Co., Ltd., on the understanding that a separate company should be formed to carry out the work. The Cambridge Electric Supply Co., Ltd., was therefore constituted, with Mr Finch, honorary fellow of Queens' College, as chairman and Sir Charles Parsons as managing director. In consideration of a payment of

£2,040, the Parliamentary powers of the Town were transferred to this Company in 1892, Parsons remaining one of its Directors for many years. The Company erected a power station in Thompson's Lane, on the banks of the Cam, and this station went into service on 19 November 1892 with 230 lamps of 10 c.p. each connected to the mains. The initial equipment of the station consisted of two Lancashire boilers working at 140 lb. pressure, and three turbo-alternators each of a rated capacity of 100 K.W. at 4,800 R.P.M. These machines were quite self-contained, each unit standing on rubber blocks in a metal tray without being bolted down. They generated single-phase current at 2,000 v. and 80 cycles, and each was considered capable of supplying sufficient power for lighting 3,500 lamps of 10 c.p. The turbines, of course, were of the new radial-flow type, for the reason already given.

In designing the station, economy in the consumption of fuel had been sought by providing a superheater for the steam and a Green's economizer for heating the feed water, but by far the most important contribution to the efficiency of the plant was the addition of condensers to the turbines. Up to this time every steam turbine that had been built had exhausted its steam at atmospheric pressure, thus following the practice that was customary with the high speed engines then in use for central station work. The steam turbine, however, as Parsons fully understood, was capable of deriving much greater benefit than a reciprocating engine, from a reduction in the back pressure, and the results obtained by the introduction of condensing practice at Cambridge and at Newcastle about the same time compelled engineers to realize that the ordinary steam engine was now faced by a most formidable competitor. This fact was definitely established by careful tests carried out by the late Professor Ewing on one of the machines built for the Cambridge Station. In his report of these tests, in August 1892, Professor Ewing remarked:

The general result of the trials is to demonstrate that the condensing steam turbine is an exceptionally economical heat engine. The efficiency under comparatively small fractions of the full load is probably greater than in any steam engine, and is a feature of special interest in relation to the use of the turbine in electric lighting from central stations. Apart from the other possible applications of a peculiarly light and efficient high speed motor, the turbine dynamo in its present state, is, in my opinion, eminently well fitted for central station use, not only on account of its economy of steam under both heavy and light loads, but also on account of its exceptional lightness and compactness, its small first cost, its independence of foundations, its freedom from vibration, its steady governing, its simplicity, the ease with which it is handled, and the moderate outlay which it may be expected to require under the heads of maintenance, oil and attendance.

The expression of so comprehensively favourable an opinion by an authority of the reputation of Professor Ewing did more than anything else to counteract the belief, due to ignorance or prejudice, that Parsons' invention was nothing more than a "toy" or nothing less than a "steam-eater", epithets which had been commonly applied to it. After the publication of Ewing's report there could no longer be any doubt that the steam turbine had to be taken seriously as a new factor in Central Station practice.

It may be mentioned that, although a superheater was included in the equipment of the Cambridge Station, the use of superheated steam was abandoned after a few months. The superheater was placed in the main flue, on the up-stream side of the economizer, where the operating conditions were apparently too arduous for it. It was stated in January 1900, by Mr Barker, the engineer to the Cambridge Station, that superheated steam had not been employed there for the past $6\frac{1}{2}$ years, owing to trouble with the superheater and was not then being used in any other power station with Parsons turbines.

The next power station to employ steam turbines was that of Scarborough. The Scarborough Station was a new one, erected by a newly formed company, like the stations at Forth Banks and Cambridge, for no established undertaking could yet be found with sufficient enterprise to instal a turbine either to supplement or to replace its existing reciprocating machinery. The responsibility of the turbines in the three stations mentioned was absolute, both as regards reliability of supply and economy of operation, for there was no other kind of machinery alongside them

which could be resorted to in case of difficulty. It is satisfactory
to record that the courage of all the three companies, who thus
trusted their entire technical fortunes to the turbine, was fully
justified by the results. Not only did they give an irreproachable
service, but they were all financially successful from the start
which is more than could be said of many of their contemporaries.

The Corporation of Scarborough had obtained a Provisional
Order to enable them to establish an electrical undertaking as
early as 1883, but as in so many instances, they took no further
steps towards giving a supply. They ultimately decided to allow
the work to be carried out by private enterprise, and, in 1891
they agreed to transfer their powers under the Order to the
Scarborough Electric Light Co., which was formed, with an
authorized capital of £50,000, to provide a service of electricity
in the town. According to the terms of the agreement, the
Corporation had the right to purchase the Company's undertaking,
as a going concern, at the end of 21 years. Alternatively, it could
be purchased at the end of 32 years, or every five years thereafter,
on the terms of the Electric Lighting Acts, i.e. without consideration either of goodwill or of earning capacity. Current was to be
sold at a maximum price of 7*d*. per K.W.H. for private lighting, and
at 6*d*. for street lighting, these prices to be reduced as soon as the
Company should be earning a dividend of 8 % on its capital.

The Company purchased a site two acres in extent and conveniently served by a railway siding, on the western outskirts of
the borough, and invited tenders for the plant required. The whole
contract, including boilers, generating units and condensing
equipment, was awarded to Messrs C. A. Parsons and Co., Ltd.,
who submitted the lowest price and guaranteed the lowest consumption of fuel at all loads. Half the land acquired for the works
was used to make a condensing pond, 4 ft. in depth, holding about
half a million gallons of water, as Messrs Parsons decided to
follow the practice inaugurated at Cambridge and supply turbines
to operate under condensing conditions. On the other half was
erected a power station measuring 60 ft. by 70 ft. in plan and
provided with a chimney of 7 ft. internal diameter at the base and
130 ft. high. The Scarborough Station went into service in

September 1893, with two Lancashire boilers and two Parsons turbo-alternators having a total capacity of 240 K.W. The boilers were 28 ft. long by 7 ft. 6 in. diameter, and worked at a pressure of 140 lb. per sq. in. A Sykes superheater was placed in the main flue behind the boilers, and a system of dampers was so arranged that, besides being operated in the ordinary way, the superheater could either be by-passed altogether by the flue gases, or it could be subjected to the direct action of the flames emerging from the ends of the boiler flues. The boilers were fed by a pair of Worthington pumps which exhausted into a Berryman feed heater. The latter could also be supplied with exhaust steam from the turbines if desired.

The engine-room, which was 60 ft. long by 34 ft. wide, was considered large enough to hold 7 turbo-alternators, of an aggregate capacity of 1,500 K.W. without overcrowding. The initial equipment consisted of two units, each of 120 K.W. at 4,800 R.P.M. generating single-phase current at 2,000 v. and 80 cycles per second, and carrying their exciters on the ends of the alternator shafts. The turbines were, of course, of the radial flow type, as installed at Cambridge, because Parsons was debarred at the time from constructing parallel flow machines, for the reason already explained. Each turbine was provided with its own jet condensing plant, but could be arranged to exhaust into the condenser of the other turbine, or into the Berryman heater. The condensing plants stood in pits alongside the walls of the engine-room. Each plant was self-contained, the condenser being fitted with a compound air-pump above and a separate water discharge pump below. These pumps were both driven directly by a rod attached to the crosshead of a small vertical tandem compound engine, with cylinders 4·5 in. and 7 in. in diameter by 8 in. stroke, which surmounted the whole apparatus.

Steam was supplied to the turbines by a cast-iron ring main, with bends of solid drawn copper. Voltage was maintained constant at all loads by a system of electrical control of the steam admission. Steam was caused to enter the turbine in a succession of gusts, due to the mechanical opening and closing of the governor valve about once in every 28 revolutions of the turbine shaft, or

more than 170 times per minute. The duration of each gust was
determined by a steam relay, which was controlled, in turn, by the
position of a plunger in a solenoid connected as a shunt across the
field magnets of the alternator. This method of governing proved
most satisfactory. There was no hunting, and the freedom from
friction, due to the apparatus being in constant motion, was
practically perfect. One of the turbines built for the Scarborough
Station was carefully tested by Professor Sir A. B. W. Kennedy in
March 1893, the machine being coupled to a continuous current
dynamo in the place of its alternator, in order that measurements
of electrical power might be made with greater precision than was
then possible with alternating current machinery. The results of
the tests, with slightly superheated steam, showed a consumption
of 27·9 lb. per K.W.H. at full load, 31·2 lb. at half load, and 44·1 lb.
at one quarter load.

No attempt was made to run the alternators in parallel, as it
was considered that, in view of the flatness of the efficiency curve,
the gain in economy by parallel running was not worth bothering
about. The switchboard included panels for 3 alternators and 4
feeders, with provision for extensions to serve 7 alternators and 12
feeders. Each of the generator panels was provided with a voltmeter, an ammeter, and a Thomson recording wattmeter, regis-
tering the output in K.W.H. The only instrument on the feeder
panels was an ammeter on each. The 2,000 v. feeders were india-
rubber cables drawn through cast-iron pipes. They delivered the
current to transformers in cast-iron street boxes generally
located below the pavements, though sometimes the transformers
were placed on the consumers' premises. Every transformer was
equipped with a Cardew earthing device, and provided with H.T.
and L.T. fuses in compartments with separate water-tight covers
in the top of the transformer box. The low-tension current, at
100 v., was distributed also by I.R. cables in cast-iron pipes. All
the pipe joints were run in with lead to make the system water-
tight. To prevent trouble from moisture, air, previously dried by
passing to and fro over trays containing chloride of calcium, was
pumped into the pipe-system under a pressure of 2 in. water
gauge at the power station, and only allowed to escape at the

extreme ends of the pipe lines. The air-pressure was maintained by a Roots blower driven by an auxiliary steam engine.

For the first three months after going into commission, the Scarborough Station only gave a supply during the hours of darkness, but at the end of December 1893 the service was made continuous, day and night. There were then the equivalent of about 4,000 lamps of 8 c.p. connected to the mains, but the load grew rapidly and a third turbo-alternator of 75 k.w. capacity was erected in 1894. This was followed by a further unit of 150 k.w. in 1895 and another of 500 k.w. capacity at 2,400 r.p.m. in 1900. The latter machine, when working with saturated steam at 126 lb. pressure at the stop-valve and exhausting into a vacuum of 26·8 in., showed a steam consumption of 22·7 lb. per k.w.h. at an output of 529 k.w. At a load of 258 k.w., the consumption was 26·4 lb. per k.w.h. with saturated steam at 128 lb. pressure and a vacuum of 27·65 in.

The next central station to take advantage of the new form of prime mover was that of the Municipality of Portsmouth, which was inaugurated on 6 June 1894 with one Parsons steam turbo-alternator of 150 k.w. and two slow-speed reciprocating units of 212 k.w. capacity each. The Portsmouth Station was the first municipally owned plant to employ a steam turbine, and was the first central station of any kind to employ turbine machinery running in parallel with reciprocating sets. The turbine, which was of the radial flow type, drove an alternator generating single-phase current at 2,000 v. and 50 cycles, the speed being 3,000 r.p.m. The set stood on rubber blocks which rested on the concrete foundations, no holding down bolts at all being used. It exhausted into a jet condenser. Below the condenser was the water-pump, and above it a compound air-pump. The pumps were worked from the cross-head of a small tandem compound engine with cylinders 4 in. and 7 in. diameter by 12 in. stroke, mounted on girders on the engine-room floor above the condenser pit. A fuller account of the Portsmouth Station was given in Chapter VI.

A striking opportunity for the turbine to demonstrate its peculiar merits was afforded by the troubles of the Metropolitan

Electric Supply Co. in 1894. The Manchester Square Station of this Company was opened in January 1890 with ten single-phase alternators of 120 K.W. capacity, directly coupled to Willans engines running at 350 R.P.M. There was no reason to anticipate any difficulty with the plant, which conformed to what was considered to be the best practice of the day. However, complaints of vibration were received from neighbouring householders almost from the day the station was started. These grew more and more insistent, and by 1892 it had become quite evident that something had to be done to remedy matters. Investigations showed that, 10 or 12 ft. below the engine-room level was a bed of spongy clay or mud, some 23 ft. in thickness, on which the foundations of the engines practically rested. It was decided by Mr Frank Bailey, the Chief Engineer of the Metropolitan Company, to try supporting the continuous concrete bed which formed the foundation for the engines, by cast-iron columns reaching right down to the firm clay beneath. The columns were filled with concrete, and one was placed between every pair of the ten engines which the station then contained. The work was completed in 1893 at the cost of £2,000, but complaints of vibration continued and further efforts to ameliorate matters by the addition of sloping struts to prevent any motion of the tops of the columns were also fruitless. Legal proceedings were taken against the company by some neighbouring householders, and although the judge agreed that everything that experts could suggest had been done to remove the cause of complaint, he found, nevertheless, that a serious nuisance persisted, and in April 1894 an injunction was granted against the company, with a suspension of three months. The engines which were causing the trouble were two-crank compound Willans engines of 200 I.H.P. each, running at 350 R.P.M., and similar to hundreds of others running with perfect satisfaction elsewhere. Each line of their reciprocating parts weighed 420 lb. and had a stroke of 9 in. Extra flywheels, carried on the outer ends of the crankshafts, had been fitted in the hope that they would make some improvement, but as the only effect of them was to bring about breakages of the crankshafts, they were taken off again.

The situation had become extremely serious, for there appeared to be no way of complying with the terms of the injunction, and everything pointed to the likelihood of the station having to be shut down altogether. At this juncture, Mr Bailey, who was of course aware of the behaviour of the turbines at Cambridge and elsewhere, where they were running without even being bolted down, approached Sir Charles Parsons in the hope that he could help him out of his difficulties. The outcome was that Parsons undertook to build a turbo-alternator of 350 K.W. capacity at 3,000 R.P.M.—a unit of more than double the power of anything of the kind previously constructed—and this machine, which was delivered and running before the contract time, demonstrated that the turbine could save the station. Other sets of the same capacity were installed as rapidly as possible, and in a short time both the Manchester Square and Sardinia Street Stations were completely equipped with the new type of generating machinery. All these turbines were of the parallel flow type, for Parsons had recovered the use of his patents in 1894, whereupon he immediately abandoned further development of the radial flow principle.

The next opportunity for any striking advance in turbo-generating machinery was afforded in 1900 by the Municipal Authorities of the City of Elberfeld, in Germany, who, on the advice of Sir W. H. Lindley their consulting engineer, placed an order with Messrs Parsons for a pair of single-phase turbo-alternators, each with a capacity of 1,000 K.W. at 1,500 R.P.M. These turbines were not only the largest in any power station in the world, but were also the first to be built with two cylinders in tandem. Their steam consumption of 18·22 lb. per K.W.H. when condensing, which was attested by an International Board of Engineers appointed by the German Authorities, was rightly considered a remarkable performance at the time. The manufacture of turbine machinery was immediately taken up on the Continent by Messrs Brown, Boveri and Co., working under licence from Messrs Parsons, and was developed by the Swiss firm with great success. It may be mentioned that the American rights in the Parsons patents had already been acquired by the Westinghouse Company in 1896.

A considerable improvement in the economy of steam turbines was made in 1903 by the introduction of the Parsons "Vacuum Augmentor". The function of this device was to extract the air from the condenser by means of a steam-jet, and to deliver it in a much less rarefied condition, to the ordinary air-pump which then could handle it more efficiently. The effect of the Augmentor was to improve the vacuum by an inch or more, with a corresponding reduction of about 5 % in the steam consumption of the turbine, after allowing for the steam used in the jet. The Augmentor was the prototype of the modern steam-jet air-ejector now universally used for the extraction of air from condensers, the only difference being that in the modern apparatus the principle is carried to its logical conclusion by eliminating the air-pump altogether and replacing it by one or more additional steam-jets.

A further advance in power station practice was made in 1905 when the first turbo-alternators to generate at 11,000 v. were supplied to the Kent Electric Power Co. for their Frindsbury Station. These machines had a capacity of 1,500 K.W. at 1,500 R.P.M. Turbo-generators continued to increase in output, and in speed for the larger outputs, and in 1908 some very efficient tandem machines of 6,000 K.W. capacity at 1,000 R.P.M. were installed at Lots Road and elsewhere. All previous achievements as regards size and efficiency were, however, once more broken by Messrs Parsons in 1912 by the construction of a turbo-alternator of 25,000 K.W. at 750 R.P.M. for the Fisk Street Station, in Chicago. The turbine was the first to be built of the now familiar tandem type with double-flow L.P. cylinder, and it established another record for efficiency with a steam consumption of 10·45 lb. per K.W.H. It is worthy of remark that this unit, although over 26 years old, is still giving good service. It did not hold the record for efficiency very long, as its performance was beaten by that of some 11,000 K.W. 2,400 R.P.M. turbo-alternators supplied to the Carville Power Station in 1914 and after, the average steam consumption at full load of five of these machines being only 10·05 lb. per K.W.H.

By this time, and indeed for several years before it, the turbine had become established as the only prime mover for power stations of any importance, and its subsequent development belongs to more modern history than this book is concerned with. Certain later steps, however, may be just referred to, such as the progressive heating of the feed-water by partially expanded steam, which was first done at the Blaydon Burn Power Station in 1916; the use of steam at 450 lb. pressure, and its re-superheating during expansion, at North Tees in 1918; the first employment of a closed ventilating circuit for the alternator with a gilled-tube cooler for extracting the heat from the ventilating air, at Blaydon Burn in 1919; the installation of a 40,000 K.W. cross-compound turbine working on a re-superheating cycle, at Barking in 1922; the first turbo-alternator to generate directly at 33,000–36,000 V. which went into commission at the Brimsdown Station in 1928; and the 50,000 K.W. re-superheating units at Dunston in 1930, machines which still probably hold the record for efficiency with a heat consumption of only 9,280 B.T.U. per K.W.H.; and have placed the Dunston Station at the head of all the Power Stations in Great Britain as regards thermal efficiency. The largest turbo-alternator in this country is a machine of 105,000 K.W. capacity installed at Battersea in 1935, though considerably larger units are in service in the United States.

CHAPTER XI

LEGISLATION AFFECTING THE ELECTRICAL INDUSTRY

Our Acts our angels are, or good or ill,
Our fatal shadows that walk by us still.

J. FLETCHER

THE introduction of electricity as a practical means of illumination raised many questions as to the legal position of parties desirous of giving or obtaining a public supply. Gas and water were already being supplied under Statutory Powers, and it was therefore natural that those seeking to establish electrical undertakings should wish to make their position secure by obtaining rights of some similar kind. During the year 1878 there were 34 private bills presented to Parliament with the object of obtaining powers to supply electricity in various towns. Confronted with a new problem for which there was no exact precedent, the Government appointed a Select Committee to consider whether Municipal Corporations should be authorized to undertake schemes for electric lighting, and under what conditions, if at all, Gas Companies or other Companies should be empowered to do so. The Committee decided not to make recommendations in any particular cases, but to hold an enquiry, and in the light of the evidence obtained, to lay down general principles only, leaving all Bills to be dealt with by the regular Committee of the House of Commons.

The Select Committee commenced its enquiry on 31 March 1879, under the Chairmanship of Sir Lyon Playfair, F.R.S., and amongst the witnesses called were men of the eminence of Professor Tyndall, Sir William Thomson, Dr Siemens, Dr Hopkinson and others. The Report, which was issued on June 19, was not received with enthusiasm by those interested in the development of the electrical industry. After referring to the opinions of the scientific witnesses that electricity was destined to take a leading part in public and private illumination, and that in future it might be used to transmit power as well as light over considerable distances, the Report continued: "In considering how far the Legislature should intervene in the present condition of electric lighting, the Committee would observe generally that, in a system which is developing with remarkable rapidity, it would be lamentable if there were any legislative restriction calculated to interfere with that development." If only this pious sentiment had been acted upon by Parliament, how much more healthy and vigorous would have been the growth of the industry!

No powers, it was pointed out in the Report, were required to enable large establishments such as theatres, halls or workshops to generate electricity for their own use. If Corporations and other local authorities had no power under existing statutes to take up streets for electric light cables, the Committee thought that ample powers should be given them. The time, however, had not arrived, in the opinion of the Committee, to give general power to private electric light companies to break up streets except with the consent of the local authorities, although it was desirable that the latter should have power to grant facilities to companies or private individuals to conduct experiments.

The final view of the Committee, which had such a deleterious effect on subsequent legislation, was as follows:

When the progress of invention brings a demand for facilities to transmit electricity as a source of power and light from a common centre for manufacturing and domestic purposes, then, no doubt, the public must secure compensating advantages for a monopoly of the use of the streets. As the time for this has not arrived, the Committee do not enter into this subject further in detail than to say that, in such a case, it might be expedient to give to the municipal authorities a preference during a limited period to control the distribution and use of the electric light, and, failing their acceptance of such a preference, that a monopoly given to a private company should be restricted to the short period required to remunerate them for the undertaking, with a reversionary right in the municipal authority to purchase the plant and machinery on easy terms.

Which might be interpreted as meaning that no electricity undertaking should be allowed to be established by private enterprise until the municipality had attained the limits of delay, and then only on terms that suggested the compulsory sale of the plant in a few years' time at scrap-iron prices, with no consideration for the goodwill that had been built up. Furthermore, inherent in the proposal was the idea that no undertaking, whether Municipal or otherwise, should be empowered to give a supply of electricity over an area greater than that under the jurisdiction of the local authority, so that the industry generally would consist of small parochial units.

Following the Report of the Select Committee, Mr Joseph Chamberlain, then President of the Board of Trade, introduced an Electric Lighting Bill into the House of Commons on behalf of the Government in 1882. It was a deplorable measure, even in the form in which it was finally passed, for its practical result was to bring about an almost complete paralysis of the industry until 1888, when it was amended by Parliament. To understand the motives that prompted the Government, it is necessary not only to remember the strong predilection of Mr Chamberlain, enhanced, no doubt, by his own great achievements as Mayor of Birmingham, for municipal control of all undertakings affecting the interests of the public in any area, but also to take into account the state of popular feeling at the time. Monopolies had been granted to gas and water companies, some of whom had made themselves thoroughly unpopular by what were considered excessive charges and high-handed methods. The public resented what they felt to be the exploitation of their helplessness, and there was a general determination that Electric Lighting Companies should not be granted perpetual concessions that might be open to abuse. This feeling was evidently shared by the members of the Select Committee, and considering the views expressed in their Report, the Government can, perhaps, hardly be blamed for acting as it did. It must also be remembered that, in 1882, long distance transmission of electricity was hardly dreamt of. Such generating plants as existed were exceedingly small, and there seemed nothing to suggest that it would be an uneconomical procedure for every local authority, or even every parish, to possess its own generating station and its own system of mains for the supply of such of its residents as might desire to have the new kind of light. The fault of the Government was that it looked too far ahead in trying to shield the public from the dangers of a hypothetical monopoly, and not far enough as regards the immediate requirements of the situation.

The Bill received its second reading on 18 April 1882, and, on the motion of Mr Chamberlain, it was at once referred to a Hybrid Committee, nine of whose members were to be selected by the House, and six by the Committee of Selection. On June 3 the Committee reported, their chief recommendations being that the Board of Trade should be empowered to grant licences to local authorities, or to private companies with the consent of the local authorities, to establish electricity undertakings in defined areas, these licences to be for not more than 5 years, but renewable. The Board of Trade should also be empowered to grant provisional orders for a similar purpose to local authorities, or to private companies without the consent of the local authorities, but these must be confirmed by Parliament. Both licences and provisional orders should carry with them the power to break up streets. It was further recommended that the local authorities should be empowered to purchase the undertaking of any company at the end of 15 years or at the end of every period of 5 years thereafter. The original draft of the Bill had actually envisaged compulsory purchase at the end of the first 7 years. The price to be paid would be the fair market value of the plant, with no consideration of goodwill. No overhead mains should be put up without the consent of the local authorities. Local authorities should be required to keep separate accounts of their electricity businesses, and publish them in detail for the information of the ratepayers.

While the Electric Lighting Bill was before the House of Commons, protests against its provisions were raised both by companies and local authorities. The former objected to the clause whereby their undertakings would be subject to purchase after so short a period as 15 years without consideration for goodwill, while many municipalities were indignant at the thought that the Board of Trade should be enabled to grant provisional orders for electricity undertakings to be established in their areas without their consent, even though the orders would have to be confirmed by Parliament before becoming effective. Mr Chamberlain was deaf to both parties. His belief in the advantages of municipal ownership made him determined to facilitate the acquisition of company undertakings by local authorities as soon as ever there was any profit in doing so, but on the other hand he was equally determined that no local authority, swayed either by conservatism or by regard for the interests of its own gas undertaking, should be in a position to withhold a supply of electricity from the inhabitants in its area. He wanted electricity to be accessible to everybody who desired it, but he intended that, if possible, it should be a municipal service. Whether or not he foresaw the inhibiting effect of the purchase clause, he certainly over-rated the enterprise of local authorities. A much deeper insight was shown by Sir Frederick Bramwell, who prophesied, in a letter to *The Times* on 21 July 1882, that "This Bill will deprive the general public of the benefits of electric lighting. The local authorities will not dare to embark in a comparatively unknown undertaking, and private companies will refrain from doing so with those inequitable conditions of payment on compulsory purchase staring them in the face."

The Bill was read a third time in the House of Commons on July 28. In its subsequent passage through the House of Lords an amendment was passed raising the period before compulsory purchase to 21 years, and every 7 years thereafter. This was accepted by the Commons; the Bill as amended received its final reading on 14 August 1882 and became the law of the land four days later. It gave powers to undertakers to break up streets for cable laying and authorized local authorities to raise money for electric lighting schemes of their own. Companies could proceed either by way of 7-year licences, which the Board of Trade could grant with the consent of the local authorities, or by way of provisional orders, for which the consent of the local authorities was not necessary. There was no limit to the length of the concession under a provisional order, but, as already mentioned, the local authority had the option of purchasing the undertaking at the end of 21 years on terms that might be most inequitable to the owners. It also appeared to be implicit in the Act that no undertaking could extend its system beyond the comparatively small area under the jurisdiction of the local authority, nor could any interconnection be made between the systems of different undertakers. Such good points as the Act contained included the obligation to supply all customers on equal terms, the keeping of accounts in a prescribed form and the availability of such accounts to the public, and finally the prohibition of any regulations, limiting the consumers to any particular type of lamp or apparatus for the use of electricity.

The results of the Act were such as to justify the worst fears of its critics. It is true that, just at first, there was a rush to secure provisional orders, but as soon as the people who were expected to find the money for the establishment of electrical undertakings realized the onerous nature of the purchase clause, the schemes lay still-born on the hands of their promoters. On 10 July 1884, practically 2 years after the Act had been passed, Mr Chamberlain could only say, in answer to a question in Parliament,

One hundred and twenty applications for provisional orders have been made to the Board of Trade since the passing of the Electric Lighting Act. Of these, 73 have been granted by the Board of Trade and confirmed by Parliament. There have also been the applications for licences. One, for Colchester, will be granted in the course of a few days; the remainder have not been proceeded with by the applicants. *The supply of electricity has not been commenced under any of the orders.*

A similar story was told, about a year later, in the Annual Report of the Board of Trade. Only one application had been received for a provisional order during the previous year. Even this had not been proceeded with, and since the previous Report the Board had revoked all but one of the provisional orders granted to companies during the session of 1883. Nothing, in fact, could be done under the Act. The technical and lay press were unanimous in its condemnation. *The Electrician* of 15 March 1884 remarked that "The fanatical dread of monopoly has resulted in there being no business either to monopolize or to compete for", while the

St James' Gazette stated, as late at 1886, that "All attempts to establish a system of lighting in the thoroughfares and elsewhere are crushed by the provisions of the Electric Lighting Act of 1882. The restrictions of that Act are so excessive that it is impossible to work under them." Its utter failure is sufficiently demonstrated by the reply of Sir Michael Hicks-Beach to a question in the House of Commons in 1888, when he admitted that "Since the passing of the Electric Lighting Act, 59 provisional orders and 5 licences have been granted to companies, and 15 provisional orders and 2 licences to local authorities. *The Board of Trade are not aware of any cases in which the powers obtained are now being exercised.*" Indeed, at the time of the passing of the amended Act of 1888, there was only one provisional order (Chelsea 1886) and one licence (St Austell 1885) still remaining in force.

Disastrous as the Act was upon electrical development in this country, it could not altogether stop it. The essential powers conveyed by a licence or provisional order were those for the breaking up of the streets for the laying of cables. Wherever a company could obtain the consent of the local authority to the use of overhead lines, it could go about its business without reference to the Act. This was done by the Grosvenor Gallery Company, the forerunners of the London Electric Supply Corporation, while the latter company also kept clear of the Act by bringing their cables from Deptford to London along the railway lines which were private property. Brighton obtained its supply by overhead lines, and several other places avoided the Act in the same way. It was, furthermore, maintained by some municipalities that they possessed the legal rights under their Charters to break up their own streets for their own purposes, and that they had the power to give to a private company the benefit of those rights. At any rate central stations were established at Hastings and Eastbourne, and cables were laid in the streets with the assent of the local authorities, and the business of electricity supply was carried on without reference to the Act of Parliament or the Board of Trade. Such things, however, could only be done where the local authority was willing to facilitate the objects of the company. When the authority was not so complaisant, the company had no option but to proceed by means of licence or provisional order, and the risks involved in coming under the Act were too great to be taken.

It must not be imagined that those interested in the development of electricity supply were content with the situation produced by the Act of 1882. In November 1884 an influential deputation, comprising Sir Frederick Bramwell, Colonel R. E. B. Crompton, Robert Hammond, J. S. Forbes and other notabilities of the electrical world, waited upon Mr Joseph Chamberlain at the Board of Trade, to urge the repeal of the 21 years' purchase clause. The author of the Act admitted that some modification might be desirable, and upon his suggestion a committee was formed to draw up amendments which might be taken as representing the views of the industry as a whole. The outcome of this work was a Bill introduced by Lord Rayleigh in 1886, the main feature of which was the abolition of the objectionable purchase clause, and the conferment on electrical enterprises of the same security of tenure as was enjoyed by the gas undertakings. Another Bill, introduced by Viscount Bury, provided that the purchase period should be extended to 24 years, and that electrical undertakings compulsorily acquired by the local authorities should be paid for in accordance with their value as going concerns. Confronted by these two measures, the Government brought in a Bill of its own, which was presented to the House of Lords by Lord Houghton on behalf of the Board of Trade. The three Bills were referred to a Committee of the House of Lords, who threw out the first two, and recommended that the Government Bill as amended, should be proceeded with. In spite of the obviously urgent need for action, the Government felt themselves unable to find time for its passage, either that year or the next, but it was re-introduced by Lord Thurlow in 1888, and it passed into law on June 28 with the title of Electric Lighting Act 1888.

The new Act was a short one, consisting of five clauses only. It introduced the principle that the consent of the local authority should be obtained before the granting of a provisional order to any company desirous of establishing an electric light undertaking in the area of the authority, although the Board of Trade was given the power to override an unreasonable refusal of the local authority. It also empowered the Board to vary the terms of sale of the undertaking in such a manner as might be agreed upon between the company and the authority, but the most important provision was to extend the period of security of tenure of the companies from 21 to 42 years, and to increase the length of the optional intervals from 7 to 10 years. It was thus nothing more than an amending Act, but it removed the principal objections to the original Act, and thus gave an immediate impulse to the industry.

The large number of stations put into service in the early nineties afforded evidence of the relief brought about by the passing of the Act of 1888. Yet it provided no answer to many important questions that were shortly to become of interest. Undertakers, for example, were still left in doubt as to whether a provisional order conferred a monopoly on them, or whether it might not be the policy of the Board of Trade to permit, or even to foster, competition within their areas. Again, it was not clear what was to happen under the purchase clause, when a company had its power station in one area and its business in another. The undertaking of the London Electric Supply Corporation, which included a station at Deptford designed to supply current to a number of districts in London, was a case in point. Similar instances would undoubtedly recur, for it could hardly be assumed that no central stations should be permitted to supply electricity beyond the confines of their own parishes.

One consequence of the Act was a greatly increased activity on the part of the companies desirous of extending their businesses in the Metropolitan area. Early in 1889 the Board of Trade issued a notification that henceforward no more licences would be granted for large districts in London, but that all undertakers must proceed by way of provisional orders, which would require Parliamentary sanction to become valid. The next step of the Board was to hold a public enquiry into the various applications for provisional orders for the Metropolis. The enquiry was opened by Major Marindin on 3 April 1889, at the Westminster Town Hall, and the proceedings lasted for 18 sessions. Its extent and importance may be realized from the fact that the eight companies interested in the electricity supply of London, two railway companies, and no less than 13 local authorities were all represented by counsel or deputies. Evidence was taken from such distinguished pioneers of the industry as Lord Crawford, Sir John Pender, Sir William Thomson, Professors William Crookes and George Forbes, Dr John Hopkinson, Dr J. A. Fleming, General C. E. Webber, Messrs W. H. Preece, S. Z. de Ferranti, Alexander Siemens, T. O. Callendar, J. S. Raworth, Robert Hammond, James Swinburne, Latimer Clark and Killingworth Hedges, who were examined and cross-examined on behalf of the parties concerned.

Major Marindin's report was issued on 18 May. It was recognized at once as a masterly and impartial document, fair both as to the claims of rival companies and the rights of local authorities, and considerate of the needs of the ordinary citizen. It was wisest, he thought, to give a fair scope to all proposed systems of supply which appeared capable of being worked, so that the one found by experience to be the best might eventually be adopted generally. The effect of the Report was to lay down the lines upon which the development of electricity supply in London might reasonably proceed. The main conclusions were as follows: (1) The possibility of obtaining a supply of electricity should be within the reach of all, and the mere objection of a local authority to the entrance of a company into its area should not be sufficient to exclude it unless the authority gave evidence that it had the intention of obtaining a provisional order for itself. (2) The whole of the provisional orders for London should be, as far as possible, identical as to conditions of supply, compulsory powers, breaking-up of streets, and especially as to price chargeable for current. The maximum statutory price should not be lower than 8*d.*, although current was already being sold at less than this in the West End, where two companies were in active competition. (3) Both companies and local authorities should work under statutory obligations. (4) There was no objection to a company having powers over a large area provided that it had sufficient capital to carry out its obligations. (5) Competition in electricity supply was desirable in principle. In cases, however, where the local authority objected to any company at all, only one company should be admitted into the area. When the authority was willing that two or more companies should work in its area, the number should be limited to two, and in such a case both should not supply alternating current because of its unsuitability for motors. (6) All overhead lines should be removed within two years of a provisional order being obtained, and companies should not be permitted to invade other areas by means of overhead lines. (7) Electric cables should be laid in existing, or future, subways; and when no subway was available, the conduits should be of sufficient capacity to take the mains of companies competing in the area. Local authorities should have power to sustain reasonable objections to the site of a power station.

Another important part of the Report defined the respective areas of Supply of the London Electric Supply Corporation, the Metropolitan Electric Supply Co., the House-to-House Electric Light Supply Co., the Notting Hill Electric Lighting Co., the Kensington and Knightsbridge Electric Lighting Co., the Chelsea Electricity Supply Co., and the Westminster Electric Supply Corporation, these being the only supply companies then in the London district.

The powers reserved to the Board of Trade by the Act of 1888, under which the Board could grant a provisional order to a company in spite of the opposition of a local authority when the consent of the latter was being unreasonably refused, were very rarely acted upon. Such a case, however, did occur in 1894 in connection with the Borough of Guildford. The Holloway Electricity Supply Co., Ltd., were desirous of establishing an undertaking in that town, but the corporation would only consent on the condition that the provisional order contained a clause empowering them to purchase the undertaking within six months of the expiry of ten years after the granting of the order, or within six months of the expiry of every subsequent period of five years. In fixing the price, no account was to be taken of the commercial value of the concern, but the terms of purchase were to be those set out in the Act.

The company found itself unable to accept such conditions and the granting of the order was consequently opposed by the corporation. The Board, thereupon, arranged a conference at Whitehall with the promoters and the city officials, but the latter were obdurate, maintaining that they wished to keep the electricity rights within their own hands, although they were not prepared to apply for a provisional order for themselves, nor would they place any limit on the time which might elapse before they would make such application. They would, however, agree to withdraw their opposition to the company's enterprise provided that their conditions as to purchase were granted. The Board refused to accept such terms, holding that they were contrary to the spirit of the Act of 1888. Subsequently the corporation suggested that the purchase period should be fixed at 30 years instead of the 42 years envisaged by the Act, but the company would not agree to this as there was no corresponding concession on the part of the corporation. The Board then proposed, as a compromise, that the purchase period should be 35 years, which the company accepted. The consent of the corporation to an order on these terms was asked for, and, being refused, their consent was dispensed with, and the order was granted by the Board of Trade.

In 1897 a number of Private Bills were presented to Parliament seeking powers to do things that had never been contemplated by the existing Acts. The Chelsea Electricity Supply Co., for example, wanted amongst other things, powers for the compulsory purchase of land for power station purposes; the Metropolitan Company wanted to lay mains between their area and a new power station outside it altogether; the Central Electric Supply Co., which had no supply area of its own, sought powers for the compulsory purchase of a power station site, and authority to lay mains to serve authorized undertakers only; and finally the promoters of the General Power Distribution Co. desired sanction for a scheme to supply electricity over an area of some 2,000 sq. miles in the Midlands, with or without the consent of the local authorities in the area. The principles involved in these Bills were so novel and important that a Joint Committee of both Houses was appointed to advise the Government on questions of policy, and although the Chelsea and Metropolitan Companies got the powers they desired without much delay, it was not until 1903 that any attempt was made by the Government to give legislative effect to the recommendations of the Committee. A Bill for this purpose passed the House of Lords in 1904, but was afterwards dropped. In the 1905 and 1906 sessions two other Bills concerning electricity supply were introduced, but only to meet with the same fate. The measure which eventually passed into law was the Electric Lighting Act of 1909. Under this Act the Board of Trade could enable any authorized undertaker to acquire land compulsorily for a generating station, whether inside or outside the area of supply; the Board could also give authority for the breaking up of roads when this was necessary for a supply of electricity to be brought into an area from an outside generating station, and the Board could furthermore authorize any company or local authority to supply electricity in bulk if it thought fit. The Act provided, too, that the powers of purchase vested in local authorities should cover generating stations etc., supplying the area of the authority from outside.

Meanwhile, considerable advances had been made by means of private legislation. In 1900, Private Bills seeking powers to supply electricity over large areas were introduced by a number of companies. In spite of strong and organized opposition from municipal authorities generally, the Select Committee to which the Bills were referred, after nine weeks' investigation, approved the principle of bulk supply, subject to reasonable safe-guards for the rights of the local authorities. As a consequence, parliamentary powers were conferred on the County of Durham Electric Power Supply Co. to supply an industrial area of 250 sq. miles on the north-east coast; to the North Metropolitan Electric Power Supply Co. an area of 325 sq. miles; to the Lanca-shire Electric Power Co. an area of 1,000 sq. miles, and to the South Wales Electric Power Distribution Co. to serve an area of 1,050 sq. miles.

In the next session, that of 1901, more Power Bills were in-troduced into Parliament. They encountered much less opposi-tion from the municipal interests, and the right to carry mains through territory not supplied was generally admitted. Royal assent was given during the year to Bills granting the Clyde Valley Electrical Power Co. an area of 735 sq. miles; the Cleveland and Durham County E.P. Co. an area of 820 sq. miles; the Derbyshire and Nottinghamshire Electric Power Co. an area of 1,570 sq. miles; and the Yorkshire Electrical Power Co. an area of 1,800 sq. miles. The Acts obtained by these so-called "Power Companies" contained no purchase clause, but granted the companies a perpetual tenure, as piecemeal purchase by the innumerable local authorities in the area would have been quite impracticable. As against this advantage, the opposition of the municipalities, however, had the unfortunate result of excluding all towns of any importance from the companies' areas of operation. Within a few years, 21 power companies were established, covering between them the rural districts of the entire country, and had the municipalities been wise enough, while retaining their distribution rights, to have encouraged the companies by purchasing power from them in bulk, it would undoubtedly have been of advantage both to the consumers and the industry as a whole.

After the Act of 1909 ten years had to elapse before Parliament again took up the question of electricity supply. This period was one of great technical developments in the power station industry, and experience during the War years of 1914–1918 had shown the necessity for some effective co-ordination between the hundreds of companies and local authorities engaged in the supply of electricity. A new Act was therefore passed in 1919, providing that a co-ordinating body, to be called the "Electricity Commissioners", should be appointed by the Board of Trade. The primary duty of the commissioners was to improve the organization of electricity supply, and this was to be done by dividing the country up into a limited number of administrative areas, in each of which would be established a "Joint Electricity Authority" representative of the interests of all authorized undertakers in that particular area. The Joint Electricity Authorities were to acquire the various generating stations and transmission systems in their respective areas, and operate them to the best advantage of the area as a whole. Another provision of the Act transferred the control of electrical matters from the Board of Trade to the Minister of Transport, to whom the Electricity Commissioners were made responsible. Much good work was done under the 1919 Act, but owing to the refusal of Parliament to grant certain compulsory powers, asked for in the original Bill, it was not found possible to achieve the extent of co-ordination between the various undertakers which was necessary for the real success of the scheme. To show how bad the conditions were, it may be mentioned that in 1921 there were, in the Greater London area alone, no less than 80 separate supply authorities with 70 different generating stations between them. In the same area there were 50 different systems of supply, operating at 24 different voltages and 10 different frequencies. The same sort of thing persisted throughout the country, thanks mainly to the short-sighted views of Mr Chamberlain, whose original Electric Lighting Act of 1882 had been devised with the object of every parish being provided with its own independent system of electricity supply. Under such circumstances, and armed with quite inadequate powers, it was hardly to be expected that the Electricity Commissioners would make much headway with their task. A further Act was passed in 1922 which slightly facilitated their efforts, but it soon became evident that something much more drastic would have to be done if ever the electricity supply of this country was to be organized on a systematic basis. Consequently, in 1925 the Government appointed a committee, under the Chairmanship of Lord Weir, "to review the national problem of the supply of electrical energy". The Weir Committee, as it was called, presented its Report the same year. It recommended a policy of concentrating the generation of electricity in a comparatively limited number of large and efficient stations, all of which would be interconnected by a high-voltage "Grid" system. The supply authorities owning these stations would then sell the whole of their output to the Board owning and operating the Grid, while they and other undertakers would buy power from the Board according to their needs. The Weir Report was adopted by the Government as the basis of the Electricity Supply Act which became law in 1926.

It is this Act of 1926 which now dominates the supply and transmission of electrical energy in Great Britain. It provided for the appointment, by the Minister of Transport, of a board of seven members, to be called the Central Electricity Board. None of the members might hold a seat in the House of Commons, nor be financially interested in any electrical manufacturing or supply company. The first duty of the Board was to consider what stations (existing, or to be constructed) should rank as "selected" stations for the generation of the electricity required in the country. It had also to provide for the interconnection of these stations with one another, and with the systems of authorized undertakers, by means of a high-tension grid into which all stations could feed. Furthermore, it had to undertake the enormous task of bringing about such a standardization of frequency as would permit of the interconnections required. The selected stations were to be operated by their owners in accordance with directions as to output, etc., received from the Board, and all the electricity generated in them was to be sold to the Board at a price to be determined by the cost of production. The electricity required by the undertaking to which the station belongs was to be re-purchased from the Board at a price which was not to be higher than that at which the undertaking could have supplied itself with the same quantity, had the Act not been in existence. The Electricity Commissioners were to be the judges in this matter. Owners of non-selected stations could be compelled by the Board to shut down their stations and take their supplies from the Grid, provided that the Electricity Commissioners (and, in case of dispute, an arbitrator appointed by the Minister of Transport) are convinced that the purchased electricity would cost less than that produced by the stations in question.

Such, in outline, are the main duties of the Central Electricity Board. To carry them out, and particularly to meet the cost of constructing the Grid, building and extending power stations, and of changing systems to standard frequency, the Board were authorized to borrow up to a total of £33,500,000, to be repaid within a maximum period of 60 years.

The first contract for the construction of the Grid was placed in January 1928 and the last of its 26,265 pylons was erected in September 1933. By the end of 1937 the Grid comprised 4,180 miles of transmission lines, of which 2,938 miles were operating at the primary pressure of 132,000 v., and the remainder at 66,000 v. or under. Associated with the Grid were 297 switching and transformer stations, with an aggregate transforming capacity of 9,695,000 K.W. At the same date there were 162 "Selected Stations" under the control of the C.E.B. In accordance with the policy of the Board to concentrate production in the most efficient stations, only 21 of the selected stations were operated for the full number of 8,760 hours, 27 ran between 6,600 and 8,760 hours, 65 between 2,400 and 6,600 hours, 42 less than 2,400 hours, and 7 were shut down altogether. There were altogether, in Great Britain, about 440 power stations belonging to authorized undertakers, but every year sees more of these shut down and the supply taken from the Grid.

APPENDIX I

THE HEAT CONSUMPTION OF POWER STATIONS

Someone, in the years gone by,
Once did something. Was it I?
Let me be, I have forgot
Whether it was I or not.

From *The Strenuous Life*

REGARDED from a technical point of view, the function of a steam power station is to produce the quantity of saleable electricity required of it for the least possible consumption of heat in the form of fuel. In designing a station many other things have, of course, to be taken into consideration, particularly the capital expenditure that can be justified in any case, but once the station has been built, the primary object of the engineer must be to reduce its fuel consumption to the minimum, since the cost of fuel is usually greater than all the other operating expenses put together.

One of the most effective steps towards securing greater efficiency in the utilization both of coal and steam, is to adopt a method whereby a continuous check can be kept upon the performance of the plant, so that every variation from the normal, whether favourable or otherwise, shall attract immediate attention. A mere record, graphic or otherwise, of the pounds of coal burnt per K.W.H., or of the steam used per K.W.H., does not afford the check required because these figures are bound to vary so greatly with the magnitude of the output per shift that they really have very little significance. The simple graphic method devised by R. H. Parsons in 1914 is therefore now commonly employed as a means of control. Its peculiar advantage lies in the fact that it provides a criterion of operation, both for the boiler-house and the turbine-room, with which the working results may be usefully compared, no matter what the output may have been during the period. Furthermore, when full advantage is taken of its possibilities, it throws a valuable light on the conditions obtaining in the plant.

To employ the method in its most elementary form, a chart is prepared, as in Fig. 1, on which the weight of fuel consumed is plotted over the corresponding output of the station, shift by shift. So long as the intervals of time are equal, it is not necessary that they should be the usual shifts of eight hours' duration; the points might be plotted for daily, weekly or monthly periods instead, but by plotting them shift by shift, the range of output is greater and the chart is thereby rendered much more informative. When a sufficient number of points have been plotted, it will be found that they lie in a fairly narrow inclined band across the chart. It is not then difficult to draw a straight line through them which will represent the average about which they lie. Once this line has been established, it can be used as a standard of operating efficiency, for it is evident that every point subsequently plotted which falls above the line indicates a performance worse than

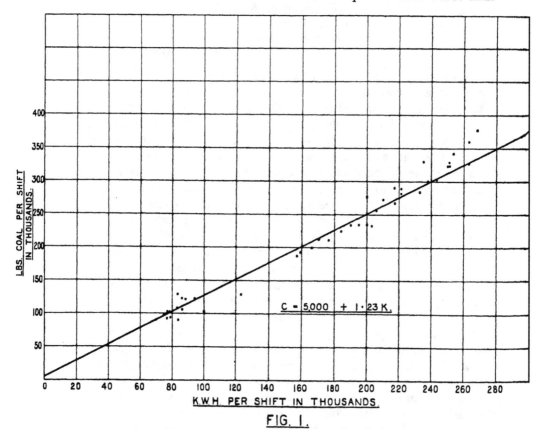

FIG. I.

the average for that particular output, and vice versa. Thus at the end of every shift there is an immediate indication of a good or bad performance, and the reason may be sought while the facts are fresh. The chart also affords a ready means of determining whether any change in operating methods, or in the nature of the fuel burnt, has been beneficial or otherwise. The effect of the change will be known with reasonable certainty at the end of the first shift after the change, for the method eliminates the disturbing influence of the magnitude of the output.

The investigation may be carried further by preparing two more charts of a similar nature on one of which the total steam consumption per shift is plotted against the output, while on the other the coal consumption is plotted against the steam consumption. Such charts are illustrated in Figs. 2 and 3. Their use enables a closer check to be kept on the operating efficiency than is possible by the use of the

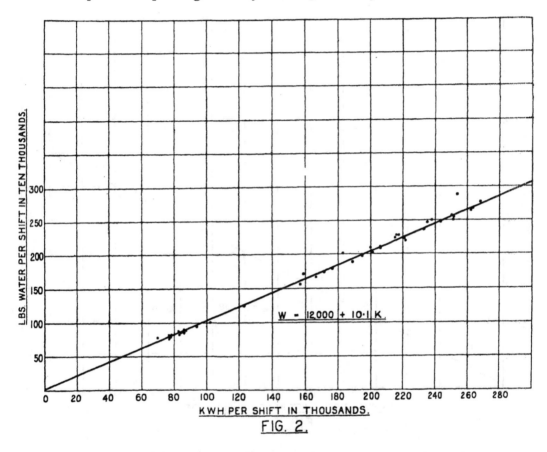

FIG. 2.

former chart alone. For example, if an excessive fuel consumption on any shift is shown by Fig. 1, reference to the corresponding point plotted on Figs. 2 and 3 will make it evident whether the fault lay with the boiler plant or in the turbine-room. It may be mentioned that the charts illustrated, represent the actual operating results for a fortnight of a large British power station.

The line on each chart may, of course, be represented by a simple equation which may be determined directly from the chart in the usual manner. Referring for example to Fig. 1, let the coal consumption per shift be denoted by C, and the output per shift by K. The value of C for any point of the line is obviously equal to the height at which the line meets the vertical through O, plus the additional height due to the slope of the line. The first of these quantities may be read directly from the vertical scale on the diagram, and is 5,000 in this particular case. It will also be seen from the diagram that when, for instance, K

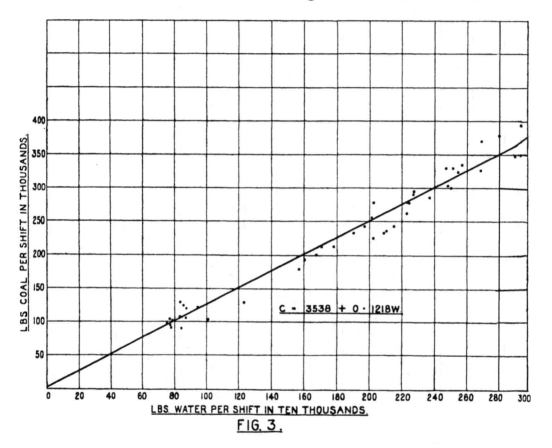

FIG. 3.

is 280,000 the corresponding value of C is 350,000. Subtracting from the latter figure the constant of 5,000, it is clear that C increases in value by 345,000 for an increase of 280,000 in the value of K, or by 1·23 for every unit increase of K. The total value of C at any point is therefore:

$$C = 5,000 + 1·23K. \qquad (1)$$

This is the equation to the line in Fig. 1, and it enables the standard coal consumption corresponding to any given output per shift to be readily determined. By dealing with Fig. 2 in an exactly similar manner, we get another equation, viz:

$$W = 12,000 + 10 \cdot 1K. \qquad (2)$$

If now we multiply each of these two equations by the numerical value of the coefficient of K in the other one, and then subtract the second of the new equations so obtained from the first, the result is to eliminate K, and to give the equation:

$$10 \cdot 1C - 1 \cdot 23W = 35,740.$$

Dividing this throughout by the coefficients of C and W in turn, we finally obtain:

$$C = 3,538 + 0 \cdot 1218W, \qquad (3)$$

and

$$W = 8 \cdot 211C - 29,057. \qquad (4)$$

These two equations are, of course, identical, but each form is useful for the purpose of bringing out some fact. It is important to notice that equation (3) must be consistent with the line in Fig. 3 and should this not prove to be the case, one or more of the lines must be adjusted until they all conform to the respective equations. The determination of equation (3) therefore provides a very useful check upon the two primary equations. A further check upon their accuracy can be obtained by using them to calculate the total weights of fuel and water which would be consumed during the period covered. During the fortnight to which the charts refer, the station from which the figures were derived actually consumed a total of 4,120 tons of coal and 74,414,550 lb. of water for an output of 7,257,800 K.W.H. It will be found that equations (1) and (2) predict these totals within an error of 1 %, and they are therefore correct within the limits of commercial accuracy.

Referring now to equation (1), this shows that 5,000 lb. of coal are burnt per shift without any corresponding production of electricity, and that an additional 1·23 lb. are consumed for every K.W.H. that is generated. The constant loss of 5,000 lb. of coal per shift represent the "stand-by" loss, or the fuel required to make up all radiation, condensation and other losses, and to maintain the plant for a shift in a state ready to carry its minimum load. These losses have to be borne whatever may be the output of the station and they are practically independent of the output. Similarly, from equation (2), the total water consumption can be regarded as being composed of two parts, namely 12,000 lb. per shift used without any corresponding output of energy, and an additional 10·1 lb. for every K.W.H. produced. As before, the constant of 12,000 lb. represents stand-by losses, leakages, etc., and remains substantially the same whatever the output of the station. Useful information concerning these constant losses of coal and steam,

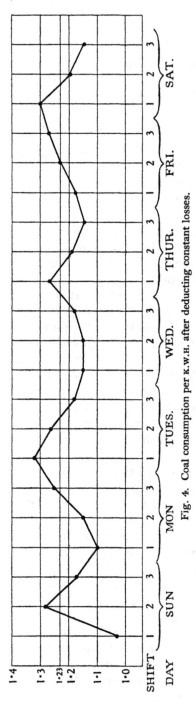

Fig. 4. Coal consumption per K.W.H. after deducting constant losses.

is afforded by equation (3). This shows that if W were zero, that is to say if no steam at all were to be used per shift there would nevertheless be 3,538 lb. of coal required to maintain the boiler-house conditions and to keep up steam. But it has already been found from equation (1) that if no K.W.H. were to be generated, 5,000 lb. of coal per shift would still be required to make up the stand-by losses. Hence the difference between these two quantities, namely 1,462 lb. per shift, must represent the coal necessary to make good the engine-room losses only, including the auxiliaries etc., if the station were running at a negligibly small load.

In addition to indicating the magnitude and location of the constant losses in the station, the equations show the efficiency to which the plant is approximating as it becomes more and more fully loaded. If, in equations (1) and (2), the value of K were taken to be very large indeed, the constant numbers would be negligible in comparison, so that the coal and steam consumptions would eventually become 1·23 lb. and 10·1 lb. respectively. By similar reasoning it is seen from equation (4) that the evaporation figure for the boiler-house is approaching more and more nearly to 8·211 lb. per lb. of coal, as the work of the boilers is increased. These limiting figures can, of course, never be obtained, but they show the efficiencies to which the station is tending when no change is made other than increasing the output. It is, however, quite true to say that any extra load given to the station can be produced for an expenditure of 1·23 lb. of coal and 10·1 lb. of steam per K.W.H. of the extra output in question, provided that the methods of operation remain the same. This deduction is a useful one to make when it is necessary to determine the net cost of carrying any particular load.

Returning to the primary use of the lines as a means of checking the station performance shift by shift, it may be considered a drawback that the diagrams do not show the order in which the various results have been obtained, and therefore do not indicate whether there is any tendency for the efficiency to improve or otherwise. This difficulty can be met in a very simple manner. In the example taken, it is evident that if the constant coal consumption of 5,000 lb. per shift be deducted from the total, and the remainder then divided by the output, the remainder will always be 1·23, no matter what the output may be, provided that the station is working at what may be called its "standard efficiency". Hence, by preparing a diagram such as Fig. 4, in which the shifts are arranged in order day by day, and plotting on it the consumption of coal per K.W.H. (calculated after the subtraction of the constant quantity) the points would all fall on the horizontal line at 1·23 lb. per K.W.H., if the standard efficiency of the station in question is being maintained. Such a diagram obviously permits the trend of efficiency to be readily observed.

An exactly analogous method can be adopted to obtain a continuous graphic record of the trend of steam consumption per K.W.H., or of evaporation per lb. of coal, and with such records a very interesting and instructive comparison can be made between the improvement in operating efficiency obtained in the engine-room and boiler-house respectively.

APPENDIX II

THE EFFECT OF LOAD FACTOR AND OUTPUT ON EFFICIENCY

Bring then these blessings to a strict account,
Make fair deductions: see to what they mount.

POPE

THE term "Load Factor" was introduced by Col. R. E. Crompton in 1891, to serve as a measure of the variation of the load on a power station over some period of time, usually a year. The annual load factor of an undertaking is defined as the ratio of the average load during a year to the peak load, or what is the same thing, as the ratio of the total yearly output to the quantity of electricity that would have been supplied if the maximum load had continued steadily throughout the year. In other words, it is the ratio of the area under the annual load curve to that of the rectangle in which the curve can be contained. From these definitions it is clear that the load factor of a station is in no way affected by the quantity of plant installed. It is purely and simply dependent upon the nature of the load itself, and would have the value of 100 % if the demand for electricity were constant, no matter what might be the capacity of the plant available or employed for production. Other terms, such as "Capacity Factor", "Running-plant Load Factor", "Use Factor" etc., have been proposed with the idea of taking the equipment of the station into consideration, but these all lack the precision which is necessary to render them really useful for purposes of comparison. As regards the capacity factor, for example, the capacity of the boiler-house may not conform very closely with that of the generating units, while there is always the question as to the proper rating of the machinery, and whether antiquated or stand-by units should be included in the capacity. The term "load factor" is open to no such objection, as it connotes something that is perfectly definite and characteristic of the duty which the station is called upon to perform.

The importance of a good load factor is obvious. A station has to be built and equipped with sufficient machinery to carry safely the peak load of the year, so that during all the time when the demand for electricity is small, machinery and cables are lying idle and earning nothing to offset their capital charges. From a commercial point of view, therefore, the profits of any undertaking and the cheapness with which it can afford to supply electricity, depend very largely upon the magnitude of its load factor. In fact, the desire for the attainment of a good load factor so dominates the commercial aspect of an undertaking, that there is sometimes a tendency to regard it as almost equally important from the technical standpoint. The belief that the consumption of fuel, or of heat, in a power station, per K.W.H. of electricity generated is related to the load factor of the station, is plausible, as it would seem to be a consequence of the knowledge that the lower the load on the generating units, the lower their efficiency becomes. The inference, however, is a fallacious one. If two similar turbo-generators both produce the same quantity of electricity in a given time, one running at a perfectly steady load, and the other developing its output in any irregular manner, the total steam consumption, and therefore the steam consumption per K.W.H., will be identical in the two cases, provided that neither machine is operated above its maximum economical rating. This is the true analogy with power stations operating at different load factors, from which it follows that the thermal efficiency of a station is unlikely to be affected by any alteration of its load factor so long as its annual output remains the same. A far greater benefit is to be expected from an increase in output than from any amelioration of the load factor, and the latter may indeed be neglected in so far as thermal efficiency is concerned.

The effects of variations in output and load factor respectively on the efficiency of a power station may be determined in the following manner. Denoting the efficiency by E, and making the assumption that this will be affected by changes in the annual output (K) and in the load factor (L), we may write;

$$dE = b\,(dK) + c\,(dL)$$

or
$$E = a + bK + cL, \tag{1}$$

in which a, b and c are constant for any particular station.

By substituting the recorded values of E, K and L for any power station for three consecutive years in the above equation, we obtain three simultaneous equations from which the numerical values of a, b and c can be calculated. An analysis of the performances of more than a score of the largest power stations in Great Britain, in this manner, was given in *The Engineer* of 28 May 1937. The result of this was to show that the average value of the coefficient c was negligibly small, and hence that the efficiency of a power station producing a given annual output, is not, in practice, appreciably affected by the load factor.

It should be noted that the type of equation (1) has no theoretical basis, but was chosen merely as being the simplest that was adequate for the purpose immediately in view. It is evident, indeed, that a linear equation could only be sufficiently accurate over a comparatively limited range, because the efficiency curve of a power station, like that of a boiler or turbine, is not a straight line.

A more fundamental examination of the effects of variations in output and load factor on the performance of a power station appeared in *The Engineer* of 20 August 1937.* It was taken as self-evident that, if the total heat consumption of any power station were plotted against the corresponding outputs for equal periods of time, the plotted points must all fall on some sort of a line, whether straight or curved, if the performances of the plant were consistent with each other. Under such circumstances, when no exact law is known, it is the standard procedure amongst experimentalists to represent the line by an equation of the form

$$H = a + bK + cK^2 + dK^3 + \ldots \tag{2}$$

the series on the right being extended to as many terms as are necessary to secure the degree of accuracy desired. Now, in the present case we may neglect all terms in which the exponents of K are even, for each of these would contribute a factor giving a positive value of H for a negative value of K, and it is manifestly absurd to imagine that a positive output of heat in a power station could coincide with a negative output of electricity. Furthermore, the terms in K^3 etc., are found to be negligibly small when their coefficients are calculated for any given power station, so that the first two terms alone remain as being of any practical significance. It therefore follows that

$$H = a + bK \tag{3}$$

gives the relationship between heat consumption and output within the limits of commercial accuracy. This equation, of course, représents a straight line and, as will be noted, is in accordance with the empirical relationship between coal consumption and output which formed the basis of the operating system described in Appendix I.

So far nothing has been said about load factor. Given the fact that the performance of a station can always be represented by a straight line, it follows either that change in load factor has no effect at all, or that different straight lines would be obtained with different load factors. The latter alternative may be examined, the assumption being made that the smaller the load factor, the higher the heat consumption for any given output, which is in accordance with what is generally believed.

* "The Analysis of Power Station Performances", by R. H. Parsons, *The Engineer*, CLXIV, 202.

Referring to Fig. 5, let the line AB represent the operating results of any station. If operation with a lower load factor would be represented by a higher line, the latter must either be parallel with AB or must intersect it. To suppose that it is parallel implies a belief that the presumed extra fuel consumption due to the lower load factor would be independent of the magnitude of the annual output, which is clearly unreasonable. On the other hand, if it is not parallel, the lines must intersect on the zero ordinate, otherwise we should have to believe that the load factor influences the constant losses of the station, which is also unreasonable. Hence we are left with the conclusion that the new conditions of operation will be represented by a line such as AC in the diagram.

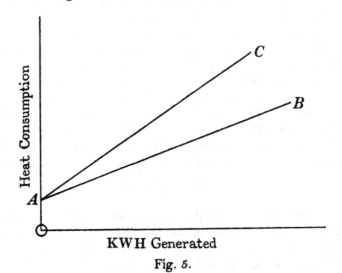

Fig. 5.

To take into account the assumed change of slope of AC with different load factors, we have to add a term to the right-hand side of equation (3), which will preserve the linearity of the equation for constant load factor, but will alter the slope of the line in accordance with the value of the load factor. There are several ways of doing this, but if we make the assumption that at zero load factor the line AC would become vertical, when an infinite heat consumption would be required to produce a vanishing output of power, the simplest form the equation can take is

$$H = a + bK + \frac{cK}{L}$$

or
$$H = a + K\left(b + \frac{c}{L}\right) \tag{4}$$

14-2

In considering the form of the equation, it should be noted that its accuracy at load factors approaching zero is of no importance because such load factors do not occur in practice. Every actual power station has a fixed amount of plant which limits the magnitude of the peak load that the station can carry. If this peak is P kilowatts, we have the relationship

$$P = \frac{K \times 100}{8760 \times L},\qquad (5)$$

and, since P is constant, a zero load factor could only coincide with a zero output. Leaving, however, zero conditions out of account, equation (5) shows that the maximum annual output of the station and the load factor are exactly proportional to each other, and this relationship, in practice, has the effect of limiting the range over which the load factor can vary, as will be seen later.

Referring back to equation (4), before this can be used for investigating the performance of any particular station, it is necessary to find the numerical values of the constants a, b and c, which are appropriate to that station. To do this we require figures of the station performance over three equal periods of time. Such figures can be obtained from the annual returns of the Electricity Commissioners, whence the output K, and the load factor L can be obtained directly, while the annual heat consumption H can be computed from the recorded figures of output and efficiency by means of the equation

$$H = \frac{3412 \times 100 \times K}{E}.$$

The application of this method of analysis may be illustrated by considering the performance of the Lots Road power station of the London Passenger Transport Board in the light of equation (4). This station, though an old one and incapable of the efficiency of modern plants, is chosen because it is a large station, unaffected by any change in equipment during the period under review.

The pertinent data for three consecutive years concerning Lots Road are given in the following table, the outputs, load factors and efficiencies being taken direct from the Commissioners' Returns, and the values of H being calculated as indicated above.

Output in millions of K.W.H.	Load factor %	Efficiency %	Heat consumption in millions of B.T.U.
K	L	E	H
365	40·4	21·15	5,888,320
396	44·6	21·42	6,307,899
413	43·9	21·45	6,569,490

By substituting the values of K, L and H for each year respectively in equation (4), we obtain three simultaneous equations, the solution of which gives us the numerical values of the constants a, b and c. Inserting these, the equation becomes

$$H = 390,946 \times 10^6 + K \left(13,792 + \frac{51,280}{L} \right). \qquad (6)$$

This equation enables us to predict what heat consumption, and consequently what thermal efficiency, the station might be expected to have when working with any assigned output and load factor, provided, of course, that it continued to "run true to form". If, for example, the output rose to 500 million K.W.H. per annum and the load factor to 50 %, an efficiency of 21·87 % would be anticipated, which may be compared with the actual operating results given in the table. Should this assumed output and load factor be attained in any future year, and found to coincide with a higher efficiency than 21·87 %, it would prove that some betterment had taken place in the meantime, either in the equipment or in the methods of operation.

Regarding changes of load factor alone, it will be seen that these have relatively small influence on the efficiency. Lots Road had, for example, an efficiency of 21·45 %, at an output of 413 millions and a load factor of 43·9 %. If the station could produce the same output with a load factor of 30 %, the efficiency would fall to 20·74 % only, or a drop of 0·71 point in the efficiency for a reduction of over 31 % in the load factor. But, as already pointed out, there is a definite limit below which the load factor cannot fall in a station with a given output. According to *Garcke's Manual of Electrical Undertakings*, the turbine equipment of Lots Road consisted of 10 sets, each rated at 15,000 K.W. Allowing two in reserve, the maximum permissible peak load would be 120,000 K.W., so that with an annual output of 413 million K.W.H. the load factor could not drop below 39·2 % in accordance with the relationship given in equation (5). Hence what efficiency this station might have at load factors below this, with the stated output, is a matter of academic interest only, for the assumed conditions would be impossible.

The initial constant in equation (6) shows that the annual no-load heat consumption of the station amounts to 390,946 millions of B.T.U. This works out to a no-load consumption of about 29,750 lb. of 12,000 B.T.U. coal per eight-hour shift. Neglecting the constant losses represented by this quantity of heat, the generating efficiency of the plant for the last year in the table would be about 22·80 %, as compared with the actual figure of 21·45 % when all losses are included.

The method of analysis described above permits the effect of variations of output and load factor to be determined for any particular station. It will be found, however, that the effect of load factor is of very little importance, as shown by the relative smallness of the coefficient c when this is evaluated. The method also proves useful in facilitating the comparison of stations having different outputs and load factors, as by means of the characteristic equation of any plant, its performance under any other conditions of output and load factor may be fairly estimated.

INDEX